Nuclear Waste Management and Legitimacy

Nuclear waste is going to be around for a long time, both during the nuclear era and the subsequent period needed to render it harmless. The legitimacy issues that arise during these generations will be irresolvable, a dilemma that society must confront in one way or another. Decades of attempts to manage nuclear waste in a legitimate manner that satisfies a range of ethical and democratic criteria have run into serious obstacles.

This book proceeds from the policy considerations, cultural understanding and ethical concerns currently associated with nuclear waste management. It examines some of the underlying, often hidden, ideas involved and explores the conceptual base for the ethics of nuclear waste. It aims at outlining a conceptual framework focused on responsibility and applying it to the nuclear waste issue, with a particular emphasis on analysing and discussing deep geological disposal.

It evolves the argument that nuclear waste management should consider ethical responsibility, including equity and rights of future generations. Thus, the waste must not be buried far below the surface of the earth and slowly vanish from human awareness.

This book is aimed at practitioners and organisations working on nuclear waste management, students of Environmental Ethics, as well as scholars with interest in the key-concepts of legitimacy and responsibility.

Mats Andrén is Professor of History of Ideas and Science, at the Department for Literature, History of Ideas, and Religion, University of Gothenburg.

Nuclear Waste Management and Legitimacy

Nihilism and Responsibility

By Mats Andrén

LONDON AND NEW YORK

First published 2012
2 Park Square, Milton Park, Abingdon, Oxon OX14 4RN

Simultaneously published in the USA and Canada
by Routledge
711 Third Avenue, New York, NY 10017

Routledge is an imprint of the Taylor & Francis Group, an informa business

British Library Cataloguing in Publication Data

Library of Congress Cataloging-in-Publication Data
Andrén, Mats.
Nuclear waste management and legitimacy : nihilism and responsibility / Mats Andrén.
p. cm.
Includes bibliographical references (p.) and index.
ISBN 978-0-415-69692-0 (hb : alk. paper) -- ISBN 978-0-203-12464-2 (eb : alk. paper) 1. Radioactive wastes--Management--Moral and ethical aspects. I. Title.
TD898.14.M35A53 2012
174′.96214838--dc23

2011037653

ISBN: 978-0-415-69692-0 (hbk)
ISBN: 978-0-203-12464-2 (ebk)

Typeset in Times New Roman
by Taylor & Francis Books

Printed and bound in Great Britain by
CPI Antony Rowe, Chippenham, Wiltshire

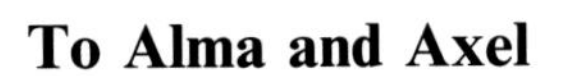

To Alma and Axel

Contents

Acknowledgements

This book is an outcome of a project entitled 'The political challenges of nuclear waste management' that I conducted together with Urban Strandberg in 2008–10. The project was generously financed by the Swedish Research Council for Environment, Agricultural Science and Spatial Planning (FORMAS). I would first of all like to thank Urban for his constructive comments on many drafts and the highly cooperative spirit he put into the project. I am very grateful for all the comments I received when presenting my nuclear waste research at seminars at the Centre for Public Sector Research at the University of Gothenburg (CEFOS) and for my colleagues in the History of Ideas and Science at the University of Gothenburg. A number of friends and colleagues have kindly and constructively commented on the drafts and ideas that I have presented. Among them are Roland Adrell, Henrik Björck, Åsa Boholm, Jan Bärmark, Mark Elam, Aant Elzinga, Johan Kärnfelt, Katarina Leppänen, Lennart Nilsson, Tore Nordenstam, Cecilia Rosengren, Barry Solomon, Sara Stendahl and Jon Wittrock. Special thanks go to Ken Schubert at Twenty-first Century Translation and Editing AB for all the effort he has made in transforming the chapters from Swedish to English. The comments of the anonymous reviewers were very helpful in finalising the manuscript, especially the thorough reading by the third reviewer.

I gratefully acknowledge the copyright holder for my article 'Uncomfortable Responsibility', *European Legacy: Towards new Paradigms* 17.1 (2012) © Routledge, Taylor and Francis Group. Chapters 4 and 5 develop ideas and arguments that appear in that article.

Preface

Human beings have always lived in communities. They are the storehouse of the systems that emerge over time – including jurisprudence, or ethics and questions about what constitutes the good life and an equitable system for ensuring the general welfare. These institutions are also bound up with the concepts that are central to modern society as it has developed since the eighteenth century.

One such concept is legitimacy, which reflects a key tenet of modern society, the imperative for citizens to rally behind the systems of laws, governance and policy-making that have been established.

The legitimacy elicited to sustain the existing order can invoke political accountability as long as the focus is on economic growth and enough citizens support it. But it is not clear whether such legitimacy also includes an ethical principle of responsibility.

Is legitimacy without ethical responsibility the wave of the future? Has it already occurred? Looking around, a fairly clear pattern emerges. The lion's share of energy around the world comes from fossil fuels. Their impact on the climate has been known for several decades. More energy is needed in developed countries to support the modern lifestyle, and in developing countries to catch up. The implications for the environment and future generations are also well known. Consumer culture is depleting the planet. Pursuing the holy grail of growth, countries improve their competitiveness while income gaps widen. The rising standard of living among the European middle class is built on the backs of cheap labour in low-wage countries or on undocumented immigrants. Immanuel Kant's admonition is more relevant than he could have imagined: 'When justice disappears, it is no longer worthwhile for men to live on earth'.[1] There is a contemporary nihilism which is intimately wedded to both inequity and environmental threats.

Equity, the environment and the safety of future generations often take a back seat to economic imperatives. Contemporary society has a propensity to stash away inequities and forget about its responsibility to the future, thereby shrinking the concept of legitimacy.

Are we condemned to legitimacy without responsibility? Are we as badly off as Harry Martinsson would have it? The 1974 Nobel Laureate in Literature published a book of poems in 1956 about a group of people who leave the earth in a spaceship named Aniara following a devastating nuclear war. As the passengers witness the destruction of the planet, the despairing narrator exclaims:

> Against all dangers can you fight,
> Fire and storm and cold and night,
> Disasters great, misfortunes small,
> But human evil withstands it all.

In the absence of unambiguous trends, the answer is no. The step that must be taken is to clearly unite responsibility and legitimacy. Economic growth must be subordinated to ethical responsibility. Responsibility must become the glue that holds society together.

Existing nuclear waste can be regarded as the legacy of a dangerous journey that humanity once embarked upon. It is a warning signal about a way of using technology that has turned out to be fraught with difficulties.

Note

1 Immanuel Kant *Die Metaphysik der Sitten,* Frankfurt am Main: Suhrkamp Verlag, 1956 & 1977 (1797), p. 453: 'wenn dir Gerechtigkeit untergeht, so hat es keinen Wert mehr, dass Menschen auf Erden leben'.

Introduction

In 1896, Henri Becquerel put some uranium salt crystals on a photographic plate wrapped up in cloth. Although the cloth did not let any light in, the plate darkened. He had discovered that uranium emits radiation. Physicists were soon able to identify various types of radiation from uranium mineral, after which Marie Curie managed to separate elements that were much more radioactive. Radium turned out to emit a million times as much radiation as the uranium atom.

Physicists realised from the very beginning that radiation was a powerful release of energy from atomic nuclei. The discipline of nuclear technology and the concept of radioactivity were born.

Nuclear power is an attractive source of energy in a number of ways. However, the waste it produces contains radiation that cannot be seen, heard or smelled. Even very small doses harm living tissue, and larger doses cause disease and death. Those who survive run a serious risk of genetic damage.

Nuclear technology and radioactivity has spawned an entire discourse about the future of the planet. What are the prospects that the human race and civilisation will survive? How do the various risks and opportunities stack up against each other? How can the benefits of nuclear technology be safely exploited and who will enjoy its rewards? How can radionuclides be isolated from the biosphere and who will manage the waste?

Nuclear technology places special demands on society. Both nuclear weapons and nuclear energy for peaceful purposes require a large measure of security and monitoring at the international level. This book focuses on nuclear waste management, which can work in democratic countries only if viewed as legitimate by the population. However, the problems are such that they can be resolved only if fundamental aspects of the modern notion of legitimacy are set aside.

Consider the concept of deep geological disposal, which casts doubt on the maxim that citizens must both participate in (individually or through their representatives) and comply with public policy decisions. Citizens who approve the construction of a repository are committing future generations to the project. Assuming that citizens of the future don't take it upon themselves to open the repository, the solution might be a reasonable one. But if they behave differently than expected due to lack of knowledge, they may have a serious problem on their hands. While such considerations also apply to many other social issues, nuclear waste is unique in that it remains hazardous for such unimaginably long periods of time. If a technology falls into disuse and current knowledge goes by the wayside, the consequences could be devastating.

Nuclear waste challenges the concept of legitimacy for both this and other reasons. Democratic societies must create legitimacy around difficult issues. None of the countries that produce nuclear energy have succeeded in that effort during the four decades that nuclear waste has been on the public policy agenda. This book posits the inability of democracies to establish such legitimacy as an explanation for the current absence of public policy decisions that can identify a solution. Efforts in that direction nowadays try to limit the foundations of legitimacy. Given that nuclear waste will be around for many decades at the very least, decisions to construct deep geological repositories must be accompanied by legitimacy that is sustainable in the long term. But how can such legitimacy be created? During the decades that a new repository is being built and waste is buried there, an accident at a nuclear plant or similar facility can quickly destroy any legitimacy that has been established. Finally, even if the repository is built and can withstand seismic shifts, it cannot be protected against the danger that somebody will one day break the seal without knowing how to handle the waste.

This book proceeds from four arguments. The first argument is that the current concept of deep geological disposal compromises the notion of legitimacy, which cannot be achieved at this point given the long-term uncertainty associated with nuclear waste.

The second argument is that the notion of legitimacy must be expanded to include an ethical dimension when dealing with issues that extend far into the future and that demand a greater sense of responsibility for succeeding generations. Nuclear waste management needs to act responsibly so as not to endanger the safety and security of future generations.

The third argument, which follows from the first two, is that that the concept of responsibility is at stake. Nuclear waste raises questions about our responsibility for the by-products of technology. Different notions of responsibility are pitted against each other. Much of current nuclear waste management is clearly responsible and includes reliable monitoring. But there are also certain tendencies towards a nihilistic approach that strives to sweep the problems under the rug. An examination of deep geological disposal, the most widely discussed solution currently under consideration, reveals that it has nihilistic features with respect to a sense of responsibility for future generations.

The fourth argument is that the current deep geological disposal solution is nihilistic. It does not meet the criterion of ensuring that waste be managed such that future generations are not harmed. The waste must not be buried far below the surface of the earth and slowly vanish from human awareness. Thus, nuclear waste should be placed in monitored, guarded intermediate storage until a fully responsible solution emerges. Until then, constant maintenance of repositories, surveillance and isolation from the biosphere, can instead serve as rituals of responsibility.

This book proceeds from the problems currently associated with nuclear waste management. It also examines some of the underlying, often hidden, ideas involved. The history of ideas is called upon in order to clarify some of the nuclear waste issues. Historical accounts of the notions of legitimacy, nihilism and responsibility set the stage for a richer understanding of them, as their meanings are always generated within specific contexts. Confusing features of the polarized debate on nuclear waste management can thus be elucidated. Difficulties associated with attempts to understand nuclear waste management confront ideas that have gained prominence in political thought since the French Revolution. Nuclear waste challenges and demands redefinitions of both legitimacy and responsibility in light of the time horizon that has opened up. The interplay between underlying ideas and the problems associated with nuclear waste charts the overall direction of the exposition.

The management of nuclear waste starts off from its existence. The three main technical solutions are reprocessing, deep geological disposal and transmutation. The nuclear waste issue is shaped by policy considerations, cultural understanding and ethical concerns. Chapter 1 frames the issue. It argues that policy is often confused with political considerations, that controversies are linguistically charged, and that the issue brings with it an ethical dilemma that reflects various myths. Special attention is given to previous research.

Legitimacy was an important concept for understanding society in the nineteenth and twentieth centuries. Constructed in the late eighteenth and early nineteenth century in the wake of the American and French revolutions, the concept captures key aspects of modern society, including the existence of generally accepted principles and a focus on governance that is based on a spirit of affinity and tolerance. While exactly what percentage of the population needed to support the existing order remained an open question, divine or historical right was no longer a justifiable claim. Legitimacy also implied that a system of governance and its policies must be constantly re-examined. While the democratic implications of legitimacy materialised only gradually and uncertainly, the concept was integral to the great advances of democracy in Europe. When it comes to the nuclear waste issue, legitimacy must now deal with the extraordinarily long-lived hazards. Nuclear waste is going to be around for a long time, both during the nuclear era and the subsequent period needed to render it harmless. The legitimacy issues that arise during these generations will be irresolvable, a dilemma that society must confront in one way or another. Decades of attempts to manage nuclear waste in a legitimate manner that satisfies a range of ethical and democratic criteria have run into serious obstacles. Chapter 2 looks at nuclear waste management against the backdrop of legitimacy concepts. A range of historical interpretations of legitimacy is presented, suggesting a general trend towards a broader concept. The perspective is based on studies from seven different countries. Each interpretation of legitimacy is shown to be problematic in its own way. In nuclear waste management, a need exists to identify a wider and more encompassing notion of legitimacy if a solution is to be found. Decisions proceed from narrow meanings of legitimacy that can bypass crucial values.

Proponents of deep geological disposal refer to ethical principles in order to focus on responsibility, including intergenerational considerations. Chapter 3 examines their concept of an ethic for nuclear waste management and argues that it is wrongly constructed and biased towards geological disposal. It states that legitimacy should not be understood as trust. Legitimacy is open to the critical dimension of ethics and loyalty for the wellbeing of both current and future generations, while trust is not. The chapter further examines correlations between legitimacy and ethics, using both Kant and contemporary social theory.

A nihilistic approach to nuclear waste management can be juxtaposed against responsibility. The most radical form of nihilism involves stashing the waste away, whereas a more moderate attitude is the refusal to assume full responsibility. The nihilistic attitude makes less strict ethical demands with respect to protecting people from radiation, whether now or in the future. Chapter 4 begins to reconstruct the responsibility concept. It initially zeroes in on the concept of nihilism, given that nuclear waste has been produced for decades in the absence of a long-term solution for managing it. The chapter shows how the discussion of nihilism has been incorporated into a critique of technology and how the concept of responsibility has been introduced as an alternative. The new political thought that arose after the late eighteenth century, primarily in relation to representative democracy, also included the concept of responsibility. The multiple levels of contemporary society make the concept more complex. The long time horizons associated with climate change, nuclear waste and other issues put responsibility in a new light such that it extends for many generations into the future. The links between the pre-war critique of technology, nuclear technology and the new responsibility concept are traced to Karl Jaspers and Hans Jonas. Hans Jonas stresses the apocalyptic background of the responsibility concept, which must be redefined to serve as a means of ensuring the survival of the human race. According to Karl-Otto Apel, the threat of environmental and nuclear disaster exhorts humanity to accept moral responsibility. The dialogue concerning a new kind of responsibility that Jonas, Apel and other philosophers conducted in the 1970s and 1980s is integral to the chapter.

Nuclear waste is a unique phenomenon in that it remains hazardous for a hundred thousand years, thereby requiring a thorough examination of the responsibility it requires. The question of responsibility can be addressed both legally, in terms of judicial sanctions, and politically, in terms of the responsibility of public officials to their constituents. As subject to the imperatives of legal and political responsibility, nuclear waste management must meet the criteria for legitimacy that exist in today's world. As a concept of moral philosophy, responsibility also encompasses more fundamental assertions. From this perspective, the ethical guidelines to which nuclear waste management must conform in order to act responsibly towards future generations become the central issue.

While Chapters 3 and 4 set out a conceptual framework, Chapter 5 applies the concept of responsibility, i.e. the conceptual framework of Jonas and Apel, to nuclear waste and deep geological disposal. The implications of devising an ethical principle for managing it are discussed and critiqued. Responsibility for nuclear waste is broken down into four components, each of which is related to the concept of deep geological disposal. The chapter argues that nuclear waste management must include a historical perspective and discusses whether deep geological disposal is a feasible solution.

Chapter 6 emphasises norms while relating nuclear waste management to a moral culture or ethical discourse. Some contextual prerequisites for constructing norms are examined. The chapter analyses deep geological disposal in light of the idea of progress, which is of crucial relevance to the issue. The conclusion of the book mentions some options for institutionalising that process as applied to nuclear waste management, while emphasising that it must be related to a broad ethical discourse.

1 The nuclear waste issue

The options

Among the many proposals that have been set forth for managing nuclear waste are disposal beneath the seabed, in the ocean, in Antarctic icecaps and in outer space. The United States and other countries chose the ocean until an international prohibition was adopted in 1975. Environmental, risk and cost considerations mitigated against all of the solutions.

Three other options have turned out to be the most feasible. The first option is almost as old as nuclear power itself. Physicists realised early on that reactors could use only a small portion (currently 5–6 per cent) of uranium's energy. Thus, reprocessing was regarded as a promising technology. Waste consists of both spent uranium and the fission products generated by the nuclear reaction. Most of the spent fuel is relatively low-level uranium whose radioactivity declines to natural levels after a hundred thousand years, while the high-level fission products of plutonium, americium and curium take a million years. Reprocessing separates plutonium from the other waste and mixes it with spent uranium in a special type of reactor. The waste that is then generated must still be isolated from the biosphere for fifty thousand years.[1]

Reprocessing seemed to be the obvious option when nuclear waste became a political issue in the 1970s. But reprocessing plants, as well as plans for their construction and associated transport facilities, stirred widespread objections. Protesters pointed to problems at the Sellafield and La Hague plants, the risk of accidents during transport and the danger that the plutonium would be used in nuclear weapons.[2] Both the United States and Sweden ended up rejecting the reprocessing approach due to fears about nuclear weapons production. Sweden decided that it would neither build a reprocessing plant nor ship the waste abroad. Not all countries were equally committed to managing their own nuclear waste. Germany, the Netherlands and other Western European countries exported waste to France and Britain, where it could be used as fuel for reprocessing plants. Eastern Europe did not regard nuclear waste as a national responsibility. The Warsaw Pact countries and Finland – as well as the Soviet republics of Armenia, Lithuania and Ukraine – exported their spent nuclear fuel to Russia. Reprocessing is currently the official policy of China, France, India, Japan, Russia and the UK.[3]

Technology began to emerge in the 1980s that would permit deep geological disposal of nuclear waste until radiation returned to natural levels. The two leading alternatives were rock salt, which is very dry and unexposed to water, or bedrock, which is stable for millions of years. Finland, which has begun to build the necessary facilities, and Sweden have the most advanced programmes for deep geological disposal. Not coincidentally, they have also taken the notion of national responsibility for nuclear waste most seriously. The leaders in the field – the Swedish Nuclear Fuel and Waste Management Co. (SKB) and its Finnish counterpart Posiva Oy – are confident that they have found the ultimate solution to nuclear waste management. The KBS-3 system is based on encapsulating waste through the use of various barriers that prevent leakage into the biosphere. The capsules must withstand seismic shifts and protect the waste from water. Because groundwater flows in bedrock and has the capacity to transport radionuclides, the system must enclose waste in material that remains leak-proof for thousands of years. The present solution is to place the waste in a copper capsule and then enclose it with bentonite clay approximately five hundred metres in the bedrock, where it will remain isolated from the biosphere for at least a hundred thousand years.[4] Apart from Finland and Sweden deep geological disposal is the current policy of Canada, Germany, the United States and many of the smaller producers of nuclear energy. It might also be the future policy of the UK.[5] The alternative of retrievability has been raised in this context. Retrievability is deep geological disposal with the possibility of bringing the waste back if needed. The rationale is that nuclear waste might be of future value. The counterargument is that final disposal avoids proliferation of nuclides and gets rid of material that otherwise might be used for weapons.

Generally speaking, deep geological disposal has been regarded as a national responsibility. But Australia, Russia and Ukraine have presented plans in recent years to charge for receiving the spent nuclear fuel of Western countries,[6] which might find the alternative more profitable than building facilities of their own. Germany and China discussed a similar arrangement back in the early 1980s.[7]

The third option is based on the recycling concept, which is much more comprehensive than reprocessing. Transmutation technology would offer a number of advantages. It could multiply the use of uranium's energy to 90 per cent. The quantity of waste would be 0.5 per cent of what it is currently, substantially reducing the immediate need for a safe and secure repository. Transmuted waste would have to be stored for only a thousand years – in other words, historical rather than geological time. Finally, the process would not separate out plutonium that could be used in nuclear weapons. But the technology requires long, costly development before it becomes feasible and reasonably safe to use. As a result, only the EU, Russia and Japan are currently working on it.[8]

The existence of nuclear waste

This much is known: nuclear waste is dangerous here and now. After half a century of nuclear energy production, the volume is significant and growing.

Nuclear waste has given rise to heated discussion, political posturing, legislation, community planning decisions and programmes for managing it. Politically speaking, these materials are special in that they remain hazardous for thousands of years.

Given that nuclear waste is generated by energy production for both peaceful and military use, calculating the total quantity is a tricky proposition. A good guess is more than half a million tonnes. The growing volume is due to a number of factors. A very small percentage of waste is attributable to the first half of the twentieth century and experimentation with X-ray technology. Large quantities were first generated when plutonium was extracted at the dawn of the US–Soviet arms race. Also important in this connection was the peaceful and commercially oriented nuclear energy production that started in the late 1950s. Once the nuclear powers had built reactors for peaceful use, other countries followed suit, leaving the world with an increasing amount of nuclear waste.

Waste products must be managed in a manner that is widely perceived as legitimate. Two possible scenarios present themselves. *The strong scenario* assumes that the use of nuclear energy will continue to grow. Nuclear power plants are being built around the world and hold out the prospect of rapid expansion in both old and newly industrialised countries. A breakthrough in transmutation technology could have the same effect. If the benefits of nuclear energy began to appear so great that it spurred major international investment despite uncertainty about its consequences, much larger quantities of waste would be generated.

The weak scenario assumes that nuclear energy will gradually be phased out and expansion plans shelved. Such a trend may develop for a variety of reasons: other sources of energy are exploited, energy consumption declines or the risk that the nuclear threshold will be passed is deemed to be too great. Countries may decide that handling such risks is too expensive or that the social costs associated with creating special institutions or establishing the necessary legitimacy are too onerous – none of which resolves the issue of how to manage the large quantities of nuclear waste that have already been generated. Furthermore, the reactors and their radioactive components would still have to be disposed of in one way or another.

The question that is relevant at this point is what the alleged need for nuclear energy entails. Arguments on both sides tend to suggest that the choice proceeds from some kind of inexorable imperative. But the real issue is one of social development, the concrete public policy decisions that are made at the national, regional and international level and that affect the interests of stakeholders among energy providers.

A country may perceive nuclear energy to be necessary for various reasons: the environment, industry, prestige or competitive advantage. In each case, the perception came to be accepted in Europe, Japan and North America during the 1950s and 1960s. A majority of the countries that have notified the International Atomic Energy Agency (IAEA) of their desire to utilise nuclear energy are most interested in power generation. Once that point of view has attracted support among the social elite, legitimacy must be established for any particular approach to managing nuclear waste.

The challenges posed by nuclear waste management demand acknowledgement that the problem exists and is constantly growing. Thus, nuclear energy and waste management are subject to many of the conditions faced by other major policy areas. In a way it resembles the welfare state, which has evolved from a clearly defined, progress-oriented project to a focus on sustainability and survival. It is definitely similar to environmental policy, which has always been about regulating the waste and emissions generated by modern industrial society. Both policy areas zero in on the management of incremental problems,[9] such as motor vehicles, whose impact on the environment is gradual but inexorable. However, no other issue has been defined and publicly debated from the vantage point of a hundred thousand years. Nuclear waste is unprecedented in the cogency that it lends to the question of long-term ethical responsibility for future generations.

Vocabulary

Controversies about nuclear waste management are linguistically charged. Certain metaphors express outrage and indignation. Formulations like 'plutonium factory' and 'risk factory' emerged when plans for a reprocessing plant were discussed in the early 1970s, and 'atomic cemetery' was used to describe proposals for deep geological disposal. Another linguistic approach strove for neutrality, borrowing economic, legal, technical and scientific terms. A phrase like 'spent nuclear fuel' lent nuclear waste a less threatening countenance. Thus, in order to understand the nuclear waste issue, some observations about the vocabulary are required.

The phrase 'nuclear power' has been around for only a few decades. The word 'waste' has been used for centuries. Nuclear power carries both positive and negative connotations. It can be regarded as an environmentally friendly source of renewable energy, or as a major risk factor and cause of insurmountable problems. The word waste was long devoid of any such dichotomies. It comes from the Latin *vastus* or deserted. *Wüste*, the German word for desert, has the same origins. But waste does not connote emptiness and wilderness alone. The Germanic languages use variations of the word *Abfall*, which long conveyed a deplorable state of affairs. In medieval Swedish, it was associated with the ravages of death. For centuries after the Middle Ages, *Abfall* often suggested contempt or disdain. In religious contexts, it referred to apostasy or a straying from the path of righteousness – as in the original sin of Adam and Eve. Luther used the word to describe his plight after having been excommunicated. Politically, the word meant that an ally withdrew its support for a country or ruler. Similarly, the transition from one political camp or ideology to another during the nineteenth century was often described in terms of *Abfall*.

The power of the negative connotations associated with the word *Abfall* made it broadly applicable. The poet Gotthold Ephraim Lessing maintained that verse suffers from *Abfall* when it fails to establish a personal voice. Others used the word to speak of cultural decline or to indict those who lived in depravity by exploiting others.[10]

Waste was not associated with garbage collection and refuse until a relatively late stage, apparently around 1880. How to dispose of rubbish, excreta and other residual products was hardly a new problem. But industrialism and modernisation, along with the growing need of cities to organise refuse collection, brought the issue to the fore. Waste was now spoken of in terms of unwanted by-products, rather than religious or political defection. It came to be associated with joint social responsibility, a task that devolved on local communities.

The modern usage of the words garbage and waste (*Müll* and *Abfall* in German) is not unambiguously negative. Two different images immediately appear to the mind's eye. According to the first image, weeds and dead leaves are composted in an isolated spot and returned to the earth. According to the second image, household waste products are hidden away where nobody has to see them any longer. Both images are easily relatable to both everyday life and proposals based on deep geological disposal of nuclear waste for at least a hundred thousand years. The concept is to put the waste someplace where it no longer needs to be seen or thought about while composting it until it returns to its original, safe radiation levels.

The images may be linked to both individual and societal concepts of waste management. They can be illustrated by means of categories such as husbandry, industrialism and mass production. Industrial society wrestled with both images. The 'out of sight, out of mind' precept stands out as nihilistic and collides with growing concern about environmental impact and the intention to act in a responsible way. The most radical form of nihilism involves stashing the waste away, whereas a more moderate attitude is the refusal to assume full responsibility. The nihilistic approach to nuclear waste management makes less strict ethical demands with respect to protecting people from radiation, whether now or in the future.

The emerging insight that even invisible waste may be harmful to life and the environment has linked waste management to perceptions about the challenges and threats posed by modern society. Thus, nuclear waste management requires a concept of responsibility that carefully considers the dangers of the material and their consequences for future generations.

The current terms for nuclear waste represent a neutral approach. Waste that cannot be used for reprocessing or nuclear weapons production is referred to as high-level waste, while other types are referred to as spent nuclear fuel, thus making a clear distinction between a worthless and a potentially valuable substance. Nevertheless, both types of waste consist of hazardous, long-lived radionuclides that must be isolated from the biosphere. Nuclear waste is usually a good umbrella term when discussing the social and philosophical issues involved, given that spent nuclear fuel in a technological context may be legally classified as high-level waste by countries that do not permit reprocessing.

An ethical dilemma

Viewed in light of the Promethean and Faustian myths, the exploitation of natural resources by industrial and post-industrial society poses an ethical dilemma. The most striking case is the waste generated by nuclear reactors around the world. Nuclear technology generates large quantities of energy, but nuclear waste will remain a hazard for at least a hundred thousand years, or until a safe method is found to dispose of it.

Myths about gifts of knowledge, wealth and power that plunge human beings into the abyss are a staple of Western civilisation. One of Plato's dialogues tells of Prometheus, who steals fire from the gods to relieve the suffering of humanity. People begin to cultivate their skills, build communities and grow prosperous. However, their new knowledge also sows division, and the gods try in vain to take the fire back. Zeus bestows political governance on the human race in the hope that it will live in peace. But Prometheus is chained to a rocky crag where his liver is eaten every day by an eagle.

Goethe's story of Faust promises ultimate redemption. God decides to reward Faust for all his effort and striving, however misdirected, and admits him to heaven. He escapes from the clutches of Mephistopheles. A key scene takes place in the protagonist's study. Mephistopheles presents Faust with his famous bargain. He will serve Faust until his death on the condition that the favour is reciprocated for all eternity. Faust will escape from the constraints of earthly life and obtain everything his heart desires. His endless thirst for knowledge will be quenched. He knows what he wants and the price he will have to pay:

> And all the good that to man's race is given,
> I will enjoy it to my being's centre,
> Through life's whole range, upward and downward sweeping,
> Their weal and woe upon my bosom heaping,
> Thus in my single self their selves all comprehending
> And with them in a common shipwreck ending.[11]

Faust signs the contract with a drop of his blood. The story has been seen as a parable of the modern human condition. The arrogance and hubris it reveals offer a perspective on certain aspects of Western civilization. Mephistopheles describes Faust as follows:

> To him has destiny a spirit given,
> That unrestrainedly still onward sweeps,
> To scale the skies long since hath striven,

And all earth's pleasures overleaps.
He shall through life's wild scenes be driven,
And through its flat unmeaningness,
I'll make him writhe and stare and stiffen,
And midst all sensual excess,
His fevered lips, with thirst all parched and riven,
Insatiably shall haunt refreshment's brink;
And had he not, himself, his soul to Satan given,
Still must he to perdition sink.[12]

The myths are about unsolvable dilemmas and irrevocable decisions, as well as ideals of knowledge, justice and progress that have spawned new epochs and exacted enormous prices. Prometheus faces a crueller fate than Faust – he is doomed to suffer for thirty thousand years. We are transported from the narrow horizon of ordinary human mortality to the eternal perspective of the gods. Whether attempts to safely dispose of nuclear waste will succeed and enjoy Faustian redemption or fail and suffer eternal Promethean retribution remains unknown.

Setting out from previous research

The question is what it means to say that nuclear waste management is perceived as legitimate. The question calls forth the whole panoply of problems associated with the notion of legitimacy. The thesis of this book is that understanding the role of legitimacy in contemporary society must be the starting point for analysing various approaches to nuclear waste management. Thus, the underlying assumptions on which a study of legitimacy is to be based must be elucidated. One solution would be to simply accept that nuclear power must be used if countries are to provide for their energy needs. In that case, the issue of legitimacy would be examined on the premise that justification must be sought for various ways of running nuclear power plants.

The crucial distinction is between the notion that nuclear energy is necessary and the understanding that nuclear waste must be managed in a legitimate manner. The responsibility of nuclear energy producers to manage waste implies that the research they sponsor will ultimately seek to vindicate the ongoing operation of nuclear power plants. However, waste management is the obvious point of departure for researchers. A paradox inherent to the contemporary discussion of nuclear waste emerges at this point. Representatives of the waste management industry in countries that plan to phase out nuclear reactors or are unsure about their future try to distinguish between the issues of waste and power generation. If a reasonable waste management solution appears likely, the nuclear power industry is in a better position to advocate for expansion or continued operation of existing reactors. Gaining acceptance for a solution would be more difficult if the future of nuclear reactors were explicitly dependent on it. The distinction between public policy debate and the objectives of research is vital. Research cannot automatically assume that nuclear energy is necessary, but it must accept the existence of nuclear waste.

Social research on nuclear waste management has been conducted ever since the mid-1970s, when the issue first became a public policy matter. The initial consensus was that deep geological disposal represented the best solution. Objections have been raised since the mid-1990s – the technological risks were unacceptable and the ethical problems were troubling. Others stressed the difficulty of reaching public consensus about how best to manage nuclear waste. There are various perspectives from which nuclear waste management can be addressed by social research. One involves democracy, public opinion and political action, another is media-oriented. One perspective is from the vantage point of groups and organisations that oppose nuclear power, another brings in the needs and priorities of local communities. The list could go on and on. The most important thing to note is that the lion's share of research proceeds from the premise that justification must be sought for various ways of running nuclear power plants. The issue of legitimacy is thus examined with such a bias.[13] The author of this book does not share that premise.

This book aligns itself with the research that goes beyond one nation or one case, looking for a general approach. There are not many such book-length examinations of nuclear waste management. A study by Andrew Blowers, David Lowry and Barry Solomon, *The International Politics of Nuclear Waste* (1991), is based on the UK but makes a thorough comparison of France, Sweden, West Germany and the United States.[14] Frans Berkhout's *Radioactive Waste: Politics and Technology* (1991) compares the UK, Sweden and West Germany.[15] Urban Strandberg and Mats Andrén's edited collection *Nuclear Waste Management in a Globalised World* (2009) includes case studies of Canada, Germany, France, India, Sweden, the UK and the United States.[16] The authors of the first volume compare and explain conflicts about nuclear waste management and conclude that trust is needed. Berkhout concludes that nuclear waste management should proceed from each country's history and culture in order to legitimately resolve the conflicts between industrial and environmental considerations. Strandberg and Andrén offer a method for identifying and analysing trends and uncertainties associated with current waste management.

However, these three books do not discuss ethical questions of responsibility. Such research is found mostly in journals, with one major exception that we will shortly examine.

The research on nuclear waste management uses ethical arguments in a variety of ways. Two articles published in the late 1970s first linked ethics and nuclear waste. Gene I. Rochlin in 1977 tried to establish criteria for safe storage of waste and concluded that risk reduction (acceptance and minimisation) was an ethically reasonable approach.[17] Such a use of ethics appeared to be fairly straightforward and uncontroversial. An article by Robert E. Goodin in 1978 regarded ethics as a critical imperative instead. He argues that traditional philosophy and orthodox economics systematically discount the future such that the risks of leakage from repositories is not taken seriously. Goodin criticises traditional utilitarian philosophy for its focus on uncertainty about whether future generations will share our values and about the impact of our decisions on them. The result of such a cost–benefit analysis is that the present is valued higher than the future because it is a more secure asset. Goodin refers to alternative models that place the interests of current and future generations on an equal footing. Fundamental to his argument is that '... uncertainty, then, implies neither that we may blithely ignore nor that we may heavily discount the distant future'.[18] For Goodin, nuclear waste exemplifies a profound type of uncertainty in that its implications extend so far into the future. He concludes that nuclear energy is justifiable only if its advocates can show that alternative sources constitute a greater long-term hazard than 'the worst possible leak of radioactive wastes'.[19]

Two ways of using ethical arguments in relation to nuclear waste were evident at an early stage: either to gain consensus or to establish a critical perspective. The dichotomy has pervaded the legitimacy discourse from the 1990s to the present, in which nuclear waste and ethics have become a common topic. An issue of *Risk Analysis* in 2000 examined the intergenerational principle as related to nuclear waste and other issues. According to some of the articles, the principle dictates that contemporary generations do not expose future ones to greater risks than those they are willing to accept for themselves. Some articles also stress that equity and welfare should be uniformly and responsibly allocated among generations. But the principle is also linked to the idea that resources may be set aside for future generations to compensate for the risks they will face, giving present generations energy while leaving future ones with risks and financial compensation.[20] The intergenerational principle can be set forth as both a source of consensus and as a means of questioning nuclear waste management programmes.

Whether used to build consensus or level criticism, ethical arguments as applied to nuclear waste management incorporate issues surrounding the future and long-term responsibility into the concept of legitimacy for the first time.

As indicated above, the consensus-oriented argument uses utilitarian ways of thinking, thereby discounting the assets and risks of future generations. The cheerleaders for the contemporary renaissance of nuclear energy maintain that it reduces consumption of fossil fuels and thereby extends their viability for both us and those who come after us. Similarly, nuclear energy is alleged to be climate-friendly due to the lower carbon dioxide emissions involved. Thus, nuclear waste is a necessary evil in the service of the common good. Any solution should be based on a cost–benefit analysis, given that both contemporary and future individuals will act rationally in accordance with economic self-interest. It follows that the costs associated with nuclear energy production should not be passed on to future generations but should be borne by those who profit from it. The utilitarianism distinction between present and future appeared prominently in the marginalist or neoclassical schools of thought that emerged in the late nineteenth century and helped shape modern economic ideas. The distinction explains capital formation as a dynamic by which certain people foresee that goods will acquire added value in the future through processing or similar means. According to this view of the human condition, some people (capitalists) possess a kind of rationality that others (workers, women, children, and so on) do not.[21] The drawbacks of utilitarianism are clear for all to see: when utility is turned into a yardstick, other value systems pale.

In its critical capacity, the ethical argument evokes the tradition from Kant to Rawls and stresses the rights of current and future generations instead of utility. Kristin Shrader-Frechette, a professor of philosophy who has championed opposition to plans for deep geological disposal in the United States for several decades, is the best example of such criticism when it comes to nuclear waste management.[22] Given her training as a mathematician and physician, she can bring assessments of scientific and technological models to bear on her ethical analysis. Her book focuses on the Yucca Mountain case. According to her, a fundamental shortcoming of the US waste management programme is that it assumes the existence of a single undisputed solution, which reinforces the subjective factor that geological and technical testing always brings with it. Studies of Yucca Mountain have used simplified models that ignore the various parameters, such as earthquakes and altered water flows, whose synergies can release radioactive substances from their capsules. Assessments of study results proceed from the answers that were sought from the beginning. The highly limited data and the fact that even evidence from several decades is a flimsy basis on which to extrapolate many thousands of years into the future are conveniently forgotten. For Shrader-Frechette, geology is a descriptive science that is particularly suited to portraying the past but cannot predict how bedrock will behave in the future.[23] Her argument concerning the ethical consequences of deep geological disposal revolves around the rights of future generations. The arguments are carefully worked out. In the first place, burying nuclear waste robs them of the ability to weigh its risks on their own terms. The present generation is relieved of the uncertainty and nuisance, while the danger that radioactivity will leak into the biosphere is kicked down the road. The very inaccessibility of deep repositories creates an iniquitous situation for generations to come.[24] In the second place, the present generation cannot predict the actions of those who will follow it. Thus Shrader-Frechette maintains that deep geological disposal magnifies the uncertainty inherent to all management of nuclear waste and extends it far into the future. The better approach is to monitor the waste until a safe means of managing it has been discovered.[25] According to her, the utilitarian arguments on which deep geological disposal are based conceal the unjust transfer of risks and uncertainty to future generations.[26]

In the following chapters I want to evoke the critical capacity of ethical arguments. This book fills a gap in the literature. It discusses the legitimacy of nuclear waste management in a very broad way, starting out from a transnational perspective. At the same time it treats legitimacy in relation to ethical questions of responsibility and nihilism. It explores the conceptual base for the ethics of nuclear waste. It aims at outlining a conceptual framework focused on responsibility and applying it to the nuclear waste issue, with a particular emphasis on analysing and discussing deep geological disposal.

Notes

1 Janne Wallenius 'Nyttiggörande eller kvittblivning – transmutation eller bara förvaring?', in Mats Andrén and Urban Strandberg *Kärnavfallets politiska utmaningar* 2005, pp. 103f.
2 Evert Vedung (ed.) 'Det högaktiva kärnavfallets väg till den rikspolitiska dagordningen', Andrén and Strandberg op. cit.
3 Per Högselius 'Spent nuclear fuel policies in historical perspective: an international comparison', *Energy Policy* 37, 2009, pp. 254–63.
4 Gunnar Gustafson 'De tekniska principerna bakom det svenska slutförvaret för använt kärnbränsla – KBS 3' Andrén and Strandberg op. cit.
5 Högselius, op. cit.
6 Jane I. Dawson and Robert G. Darst 'Russia's proposal for a global waste repository: safe, secure and environmentally just?' *Environment* 47(4), 2005; I. Holland 'Waste not want not? Australia and the politics of high-level nuclear waste' in *Australian Journal of Political Science* 37(2), 2002.
7 J. Perera 'China and Sudan want Germany's nuclear waste', *New Scientist* 107, 1991.
8 Janne Wallenius op. cit. p. 103; Urban Strandberg 'Kärnavfallsfrågan utmaningar' pp. 149f. in Andrén and Strandberg op. cit.
9 Björn Linn 'Kärnfrågor i samhällsplaneringen' in Andrén and Strandberg op. cit., p. 90.
10 Walter W. Skeat *Etymological Dictionary of the English Language*, Oxford 1910. *Svenska akademins ordbok*, 1903. *Deutsche Wörterbuch von Jacob Grimm und Wilhelm Grimm*, 1854.
11 Johann Wolfgang von Goethe *Faust*, Köln 2005 (1813), p. 62: 'Und was der ganzen Menschheit zugeteilt ist, /Will ich in meinem innern Selbst geniessen,/ Mit meinem Geist das Höchst-und Tiefste greifen,/ Ihr Wohl und Weh auf meinen Busen häufen/ Und so mein eigen Selbst zu ihrem Selbst erwietern/ Und, wie sie selbst, am End auch ich zerscheitern!'
12 Goethe op. cit. p. 64: 'Der ungebändigt immer vorwärtsdringt/ Und dessen übereiltes Streben/ Der Erde Freuden überspringt./ Den schlepp ich durch das wilde Leben, /Durch flache Unbedeutenheit,/ Er soll mir zappeln, starren, kleben,/ Und seiner Unersättlichkeit/ Soll Speis' und Trank vor giergen Lippen schweben:/ Er wird Erquickung sich umsonst erflehn,/ Und hätt er sich auch nicht dem Teufel übergeben, /Er müsste doch zy Grunde gehn!'
13 Barry D. Solomon, Mats Andrén and Urban Strandberg 'Three Decades of Social Science Research on High-Level Nuclear Waste: Achievements and Future Challenges', *Risk, Hazards and Crisis in Public Policy*, 1:4, 2010.
14 Andrew Blowers, David Lowry and Barry Solomon *The International Politics of Nuclear Waste*. London: Croom Helm 1991.
15 Frans Berkhout *Radioactive Waste: Politics and Technology*. London: Routledge 1991.
16 Urban Strandberg and Mats Andrén (eds) Special issue: Nuclear Waste Management in a Globalised World, *Journal of Risk Research* 7–8/2009. The volume has also appeared as a book: London: Routledge 2011.

17 Gene I. Rochlin 'Nuclear Waste Disposal: Two Social Criteria' in *Science* 1977, 195, pp. 23–31.
18 Robert E. Goodin 'Uncertainty as an Excuse for Cheating Our Children: The Case of Nuclear Waste' in *Policy Sciences* 1978:10, pp. 25–43, cited p. 35.
19 Goodin op. cit. p. 39.
20 Solomon *et al.* 2010. Kristin Shrader-Frechette 'Duties to Future Generations, Proxy Consent, Intra- and Intergenerational Equity: The Case of Nuclear Waste' in *Risk Analysis*, 2000:6, pp. 772–78. D. Okrent 'On Intergenerational Equity and its Clash with Intragenerational Equity and the Need for Policies to Guide the Regulation of Disposal of Wastes and Other Activities Posing very Long-Term Risks' in *Risk Analysis* 1999:5, pp. 877–902. John F. Ahearn 'Intergenerational Issues Regarding Nuclear Power, Nuclear Waste, and Nuclear Weapons' in *Risk Analysis* 2000:6, pp. 763–70.
21 Mats Andrén *Borgerskapets Marx: Eugen von Böhm-Bawerk ämbetsman och nationalekonom i sekelskiftets Wien*, Stockholm 1990, and *När den nya nationalekonomin kom till Sverige: marginalismen, den österrikiska skolan och Knut Wicksell*, Göteborg 1994.
22 See Kristin Shrader-Frechette *Burying Uncertainty: Risk and the Case against Geological Disposal of Nuclear Waste*, Berkeley: University of California Press, 1993.
23 Shrader-Frechette op. cit. chapter 4, pp. 42ff, 53f, 73f.
24 Shrader-Frechette op. cit. p. 159.
25 Shrader-Frechette op. cit. pp. 160–81.
26 Shrader-Frechette op. cit. pp. 212: ' … all of them fail in general because they presuppose that some utilitarian goal (safety, avoiding terrorism, economic efficiency) justifies extreme distributive inequalities, placing severe burdens on innocent persons in the future'.

2 Elusive legitimacy

The essence of the concept of legitimacy varies from one discipline to another. These days, sociologists and political scientists are most likely to employ the concept. Sociologists tend to lend cultural and historical dimensions to legitimacy and see it within the context of contemporary discourse.[1] Political scientists treat legitimacy either as a normative concept of the criteria by which a regime is to be judged or as the basis for empirical studies that examine the conditions under which citizens submit to a particular form of government.[2]

The question of how to manage nuclear waste in a legitimate manner is seen below in the context of developments that began long before the nuclear waste issue was aroused. The first series of developments is related to the various meanings that have been assigned to the notion of legitimacy over the past 200 years. Attempts to establish legitimacy for methods of managing nuclear waste employ those meanings in different ways and to different extents. The other series of developments is associated with the relationship between nuclear technology and the notion of legitimacy. The next section outlines the notion of legitimacy and its evolution over time, followed by an examination of its relationship with the issue of nuclear waste management.

Origins and evolution of legitimacy

By describing when and how the notion of legitimacy was first used, we can clarify the different meanings of the word. The history of the concept of legitimacy shows that it has developed under shifting conditions, requirements and stimuli. This chapter reveals seven different meanings of legitimacy: (i) legality, (ii) popular will, (iii) tradition, (iv) emotion, (v) values, (vi) best available science and technology, (vii) communication and dialogue. The varying meanings have gradually expanded into a multifaceted concept of legitimacy.

The concept emerged approximately 200 years ago. The power of a monarch, as opposed to a usurper, had previously been characterised as legal. Beginning with the American and French revolutions, legal governance was related to the will of the people. The next step was to talk about legitimacy.

The word legitimacy is largely a product of the upheavals that took place between the French Revolution of 1789 and the European revolutions of 1848. The French Revolution raised the possibility of legitimate governance based on popular support, whereas the Bourbon Restoration that followed the Napoleonic Wars emphasised the legal rights of monarchs. After 1848, legitimacy was increasingly associated with the consent of the governed through their parliamentary representatives. Thus, legitimacy has been a controversial concept from the very start.

The concept of legitimate rule goes further back. The adjective, which traces its origins to Roman law, became part of the vernacular in the early modern period and grew more common in the seventeenth and eighteenth centuries. People spoke of the legitimate succession of monarchs and legitimate governance as a manifestation of divine right, a basic covenant, natural law or human reason.

The French Revolution democratised the adjective *légitime*. Demands that the state proceed from the principle of the common good were first radicalised. Legitimate policy was assumed to coincide with the interests and wishes of the citizenry. Some Enlightenment philosophers, like Jean Jacques Rousseau in *The Social Contract* (1754), combined that idea with scathing criticism of absolute monarchy. Thus, new uses of the word legitimacy confronted monarchical principles of legitimate rule.[3]

The political pamphlets published in the summer of 1789 lent the notion of legitimacy new meanings. Deputies of the Third Estate declared a National Assembly and demanded participation to ensure legitimacy. Popular representation was now associated with the concept of a legitimate legislative body. Demands for liberty and equality were linked to the notion of legitimate power that resided with the Third Estate – only the people and their representatives could make proper policy decisions.[4]

Thus, the adjective legitimate was transformed into a noun. Legitimacy could now be expressed as an attainable object. To a certain extent, that had been the case earlier as well. Peter Burke's study of Louis XIV, the Sun King, shows that his magnificent pomp was designed to reinforce his position.[5] Thus, legitimacy is a useful theoretical concept for historical analysis. But legitimacy as a noun is more; it is the subject of performative acts by historical figures. It is not simply a theoretical concept but a protagonist in actual events.

The word legitimacy gained currency during the reaction to the revolutionary upheavals. Pamphlets during the turbulent period when Napoleon fell and the allies marched into Paris spoke of the Bourbons as a legitimate dynasty. The word held out the promise of a more peaceful future, harking back to the period before the revolution and wars. Louis XVIII, the successor to the throne, was regarded as the guarantor of a better state and good rule. French diplomat Charles de Talleyrand gave legitimacy even more exposure at the Congress of Vienna (1814–15), at which Austrian Chancellor Klemens von Metternich and others used the term to refer to legally protected dynastic succession.[6]

Discussions about the new order in France gave rise to a new meaning of legitimacy. Simonde de Sismondi maintained in 1815 that only the people could cloak a ruler with legitimacy.[7] His argument foreshadowed the post-1848 period when two key ideas emerged: the state is organised on the basis of national solidarity that eschews the concept of dynastic succession, and power proceeds from a parliament that expresses the will of the people.

The word legitimacy had three different meanings during the first few decades of its use. The first meaning was that all power proceeds from the people and the second was that it is transferred in accordance with laws of the land. The second meaning reflects an older, well-established concept of legality. The first meaning suggests that governance rests on a broader foundation than laws alone, and that the laws themselves proceed from the will of the people. Furthermore, the notion of legitimacy came to be linked with the nation-state, which serves as the institutional framework within which laws are adopted, citizens choose public officials and those officials exercise their authority. Association with the nation-state added a dimension of identity to the concept of legitimacy. The shaping of traditions, invocation of emotional affinity and assertion of values by national identities suggested additional meanings.

By no means was the concept of legitimacy limited to a single set of meanings throughout the nineteenth and twentieth centuries. Early connotations were handed down and gradually supplemented, particularly after being appropriated by the burgeoning social science disciplines at the turn of the twentieth century.

Max Weber's use of legitimacy was groundbreaking, particularly because he defined the concept in his opus *Economy and Society* and made distinctions between different types of legitimacy. According to Weber, legitimacy is a concrete reality. It is not a speculative, philosophical category but the foundation on which a particular form of governance is built.[8] For one type of legitimacy, confidence in the legality is central, which is generally associated with professional bureaucracies that have specific tasks based on discipline and knowledge. The second type of legitimacy stems from the force, credibility and sanctity of tradition. The third type of legitimacy invokes charismatic dynamism, the exemplary role of a person who manifests or brings forth a particular system of governance.[9] Weber conceives of legitimacy in terms of power in a top-down perspective.

However, Weber's use of legitimacy can also be related to a bottom-up perspective of the citizenry, which is logical given that he developed his basic sociological concepts and associated typology of power during the period when universal suffrage was adopted. Ever since, legitimacy has frequently been linked to the reasons for which citizens accept a particular system. Weber captures the fundamental aspects of the latter concept of legitimacy in a simple, structured manner. He examines the implications of the citizenry's ability to imagine the establishment of a legitimate order. The behaviour of human beings can be explained by their habit of orienting themselves towards what they regard as legitimate. Weber thickens the plot by asserting that the particular system involved shapes how citizens justify it. He identifies four different reasons for which a system may be considered legitimate. The most common reason is that the system corresponds to laws that are correctly understood, both formally and legally. The reason that emerged at the earliest point in history is that the system is consistent with tradition. The third possible reason is that the system elicits emotional allegiance. The fourth reason is that the system is rational in terms of values, which Weber associates with the notion of natural law.[10] Thus, he adds three new layers to the concept of legitimacy: tradition, emotion and values.

The concept of legitimacy acquired an additional meaning, which invoked science and technology, in the late twentieth century. An influential essay by Habermas in 1968 referred to science and technology as an ideology in the sense that they are used to impute rationality to a particular phenomenon. Science and technology are means by which legitimacy is conferred. Science, Technology and Society researchers have devoted a good deal of attention to this dynamic. Basing legitimacy on science and technology also brings various technologies face to face with each other. The imperative arises to use the best available technology, an ambiguous concept. The best technology may be the one that provides the highest return on investment or utilises resources most efficiently. This meaning of legitimacy can be incorporated into Weber's discussion of values.

The distinction between industrial and risk society that Ulrich Beck coined in 1986 illustrates an increasingly important point. According to Beck, the Western world has been entering a new phase since the 1970s in which conflicts related to risk are replacing those associated with class in traditional industrial society. Environmental, nuclear, waste management and other risks are both products of modern society and transnational in character. Key to Beck's argument is the assertion that the two kinds of societies are characterised by different grounds for legitimacy. The basis for legitimacy is the struggle against poverty, hunger and other palpable material shortages in industrial society and the minimisation of risk in risk society.[11]

Issues surrounding nuclear energy production can be examined through Beck's lens. Newly industrialised countries like India and China that are still combating poverty may be more willing than the West to take the risks associated with building nuclear reactors. On one end of the scale is the expansion of welfare, consumption and the economy, which requires major new energy resources, thereby legitimising nuclear power. On the other end of the scale is the imperative to achieve legitimacy by managing risk properly as poverty diminishes.

Another meaning of legitimacy that appeared in the late twentieth century involves communication, dialogue and the involvement of ordinary citizens in public policy. Jürgen Habermas came up with a theory of communicative action as fundamental to all social orders – particularly democratic and pluralistic societies – and legitimacy.[12] The last few decades have seen widespread interest in attracting citizens to social processes such that they can share their opinions in various ways. Swedish legislation now specifies that citizens be consulted prior to construction of new motorways or rail lines. The German states have passed a number of laws to facilitate popular participation at the local level. The core of the idea is to revitalise democracy by means of greater participation and communication. Western criticism in the 1960s and 1970s of the way democracy works in reality reflected widespread interest in that objective. Citizen involvement has become a key component of attempts to confer legitimacy on political leadership or a social project.

There follows a discussion of the nuclear waste management issue from the standpoint of legitimacy and its various meanings. Legality and the relationship between the issue and popular will are examined first. Three meanings of legitimacy added by Weber (traditions, emotions, values) are then linked to nuclear waste management. One section is devoted to science and technology as a basis for legitimacy, one to the role that communication, dialogue and deliberation have come to play.

Legitimacy and nuclear waste management

Legitimacy as a concept has acquired many different meanings since the early nineteenth century. Originally narrow in scope, the concept has broadened to reflect the growing number of criteria on which legitimacy is based. Common to all the meanings of legitimacy presented above is that it can be created within the constraints of parliamentary democracy and constitutional government. Each meaning can be related to nuclear waste management and help crystallise the issues that it raises. The trend towards a broader concept of legitimacy can be observed among some of the leading European and North American producers of nuclear energy. This does not necessarily suggest that nuclear waste management decisions are based on a broad concept of legitimacy – quite the contrary. The difficulties that have emerged are due to the inability of decision-making processes to incorporate the multitude of demands that a broad concept brings with it.

The empirical evidence on which the following analysis of the various legitimacy issues faced by nuclear waste management are based comes primarily from publications of the research project that Urban Strandberg and I conducted in 2005–11. The project looked at legitimacy and compiled case studies from France, Britain and the United States (three of the biggest producers of nuclear energy), as well as Canada, Sweden and Germany (three of the next biggest producers) and India (one of the rapidly growing large developing countries).[13] The results are presented as 'Nuclear Waste Management in a Globalised World' in a special issue of *Journal of Risk Research* (2009), which was reprinted as a book with the same title by Routledge in 2011.[14]

(i) Legality

The first and oldest meaning of legitimacy is legality, the notion that governments should rule in accordance with laws and principles of equity. The political and legal system of a country shapes the kind of legitimacy bestowed on attempts to manage nuclear waste. However, the legal framework for dealing with the issue was established during the same era in all of the countries mentioned above. Meanwhile, both the Treaty on the Non-Proliferation of Nuclear Weapons and the establishment of the IAEA preceded national legislation. The purpose of the Euratom Treaty establishing the European Atomic Energy Community was to ensure that member states would retain access to uranium fuel. But the legal management of nuclear waste was left to the individual countries.

Based on a 1976 inquiry by the British Parliament that identified nuclear waste management as public policy matter, the Department for Environment, Food and Rural Affairs was given overall responsibility and the government appointed an independent advisory body.[15] The US Congress passed the first legislation in 1982. Ever since 1989, federal laws have been complicated by Nevada legislation aimed at blocking the construction of a national repository on its territory, as reported by American geographer Barry Solomon.[16]

Complex processes emerged when national institutions developed legal mechanisms for managing nuclear waste. The progression from the Swedish Stipulation Act and Atomic Energy Act of the late 1970s to the Radiation Protection Act of 1988 and the Environmental Code of 1998 is a good example of this complexity.[17] The Environmental Code requires that environmental courts examine applications to run nuclear plants but does not say anything specific about the hazards and difficulties associated with managing nuclear waste. The ruling handed down by the Vänersborg Environmental Court in 2006 concerning an application by the Ringhals plant to continue operating and increase production found that the law was difficult to interpret in several respects. According to the court, use of the best available technology to ensure the same level of safety as a new reactor would be a reasonable approach. The ruling stressed Section 1 of the Environmental Code, which speaks of society's 'responsibility for wise management of natural resources', and questioned whether 60 per cent of the energy produced should be discharged into the ocean along with cooling water. The court also pointed out that the company did not have agreements, licences or facilities to manage waste.[18] However, the government decided that the objections were insufficient grounds for rejecting the applications and granted the requested licences. Thus, existing legislation is open to widely divergent interpretations and political manoeuvring. The legal foundation for legitimate nuclear waste management remains shaky.

The idea that public policy decisions should be preceded by a thorough inquiry pervades attempts to establish legal grounds for nuclear waste management. In discussing the German process, risk researchers Peter Hocke and Ortwin Renn argue that representative democracy was intended to work that way but that civil society was not given the opportunity to exert any influence.[19] When popular protests reject proposed nuclear waste management measures, a special kind of challenge arises.

(ii) Popular will

The second meaning of legitimacy is that the will of the people is the ultimate basis of all public policy decisions. Preliminary studies in 1972 to find a suitable repository site in Lower Saxony spurred vociferous protests from the local population, particularly landowners. Exploration was halted after four years. The choice, apparently a strategy aimed at limiting the objections to one area, fell on the community of Gorleben in 1978. Nonetheless, the protests spread to the entire country when it became known that insufficient geological studies had been conducted. Gorleben was a battleground for the West German anti-nuclear movement in the years to follow. *Gorleben soll leben* was a celebrated slogan and the title of a song by Wolf Biermann.[20]

Popular distrust of nuclear waste management, both in the present and in plans for the future, was fairly widespread during the mid-1970s. Exploratory studies in northern Ontario, Canada, ran into local resistance.[21] Plans for a reprocessing plant in Sellafield, England, aroused protests in 1977–78.[22] Popular discontent among Swedes was sparked in 1970 as a response to plans for a reprocessing plant in Tanum and climaxed with the 1980 referendum to phase out nuclear power.[23]

Legitimacy in terms of both legality and popular will has been further complicated by the 'multi-level democracy' that has emerged in recent years. Legally speaking, the nuclear waste management issue in Sweden has been closely linked to local autonomy in the sense that municipalities make crucial decisions about whether deep geological repositories will be located in their areas. Any repository to be constructed must be approved at both the national and municipal level. In other words, a *Riksdag* decision is subject to a municipal veto. Meanwhile, the question of popular will arises at both the national and municipal level. Legitimacy must be established at both levels if deep geological repositories are to be built. Studies, political debate and official reviews precede decisions at the national level. Formal meetings – as well as informal forums among politicians, the Nuclear Fuel and Waste Management Co (SKB) and citizens – are intended to create local legitimacy. The setup reflects SKB's desire to win the support of the local population for deep geological disposal, according to Swedish sociologists Mark Elam and Göran Sundqvist. Such a basis for legitimacy has gained currency among municipal representatives and political scientists in recent years. As a result, construction of a repository may win a decisive majority in the *Riksdag* but not at the municipal level, or vice versa. Similarly, legislation on nuclear technology and radiation protection that stresses national interests conflicts with the Environmental Code, which emphasises consultation with the local populace. The multitude of laws reflects and reinforces the complex relationships among the various levels of society and democracy.[24] In other words, parallel popular wills exert a simultaneous influence on events. Thus, legitimacy is a matter of how they interact and are reconciled. As the ultimate decision-making body, the Swedish government is the final arbiter of where a repository will be located.

(iii) Tradition

Managing nuclear waste from the standpoint of tradition, Weber's third meaning of legitimacy, is complex because the waste poses such a major threat to the biosphere in geological rather than historical time. Traditions of nuclear waste management have begun to emerge over the past few decades, though varying from country to country. With its combination of concealment and composting, deep geological disposal is consistent with traditional management of household and industrial waste. However, nuclear waste is a horse of a different colour. The planned repositories are also exceptional in nature. Waste has never been buried so deeply and for such a long period of time. What's more, the universality of the problem tends to override national traditions and illustrates the need of creating global legitimacy.

Nevertheless, national traditions still play a role in global trade and other international developments. The relationship between the national and international levels is particularly important to keep in mind given the globalisation of nuclear energy production in recent years, as well as the need for universal standards and regulations to manage the risk that radioactive substances will be enriched and spread.

The UK continued to focus on the development of nuclear technology for military use after the Second World War. But a political will emerged in the 1950s for the country to be a world leader in peaceful use and – soon thereafter – reprocessing technology as well.[25] Defenders of the Swedish KBS-3 solution presented it as the kind of cutting-edge technology that the country had a long tradition of developing.[26] Both the British imperial tradition and Sweden's history of pursuing modernisation are closely related to the emotional component in Weber's concept of legitimacy.

(iv) Emotion

The fourth meaning of legitimacy is emotion, which includes both pride and distrust. The pride born of the ability to produce nuclear energy was palpable in the 1960s and 1970s. It was regarded as a source of national self-confidence and strength. State-of-the-art technology would enable a country to stand tall on the world stage while creating domestic welfare and prosperity. It went without saying that the ability to manage nuclear waste was a national responsibility.

A historical paradox presents itself at this point. A key motivation behind the technology of nuclear energy production and its commercial expansion was the sense of modernity it evoked. But while the technology itself conferred legitimacy, management of the waste it generated was difficult to execute or justify.

Given that methods of managing nuclear waste have trouble inspiring popular confidence, emotion is the meaning of legitimacy that leads to the most problems. Public officials, parties and the political system in general are unlikely to instil such confidence. Contemporary democracy is based on criticism and questioning of various political standpoints. Furthermore, distrust of politicians is growing among broad sections of the population. Compelling scientific solutions for burying and encapsulating nuclear waste can generate emotional confidence but only among those who have sufficient faith in technology. Public perceptions of technology are a blend of optimism, scepticism and profound distrust. Doubts about science and technology have grown since the 1970s. Whether nuclear energy and deep geological disposal can inspire the kind of emotional confidence that legitimacy requires according to Weber is far from certain. Public opinion will take different directions over the course of the next hundred years. The 1979 accident at Three Mile Island had a major impact. Major leakage or a terrorist attack at a nuclear power plant somewhere in the world could sway popular feelings about national waste management programmes. Nevertheless, basic faith in the ability of science to solve all problems remains strong. Citizens and public officials discuss various approaches to combating global warming on the basis of scientific data. Sustainable development efforts proceed from science and technology to find more efficient ways of utilising scarce resources. Appeals to emotion should be able to draw support from that reservoir of trust.

(v) Values

Emotions are closely linked to the basis for legitimacy that Weber identified in human values. The fifth meaning of legitimacy is correspondence with explicitly formulated and generally accepted beliefs.

Many of the explicit values associated with nuclear waste are common to the various countries concerned: safe management to prevent leakage into the biosphere, minimisation of risk, democratic decision-making, the quest for sustainable development during the global warming debate of the past decade, public supervision and assurance that nuclear weapons production does not benefit. Countries outside Europe and North America have not always respected such values. For instance, India, Israel, North Korea and Iran have succumbed to the temptation to ignore the non-proliferation treaty.[27]

For the past half century, the production of nuclear energy and nuclear waste has tracked three major social values. It has heavily contributed to the energy consumption of industrialised countries and thereby to the growth of the welfare state, linking nuclear energy production to the value of equity. Nuclear energy and waste have stirred political conflict, campaigns to sway public opinion and vociferous popular protest. Thus, production, its expansion and its location have been associated with the value of democracy. Finally, the public debate about nuclear energy has reflected differing views concerning the value of technology in social development. Abiding faith, particularly within the industry itself, in the ability of technological progress to overcome daunting challenges has encountered profound scepticism among critics of nuclear energy.

But there are also important values that are correlated with national pride. Countries have regarded access to nuclear energy as being in their national interest. National enterprises have built and operated the plants. Legislation, public agencies and organisations that regulate and manage spent nuclear fuel have been viewed as national concerns for the last several decades. National pride can be an incentive to develop nuclear energy and waste management. Such motivations are increasingly prominent outside the industrialised countries of Europe and North America, as exemplified by Ram Moham and Veena Aggarwal in their study of nuclear waste management in India. Homi Bhabha, the father of the Indian nuclear energy programme, stressed the importance of making the country energy-independent. He advocated a national nuclear fuel cycle, from mining of thorium (an alternative to uranium) to reprocessing plants (currently four in number).[28] Nuclear energy production has been incorporated into national identities in various ways. In the case of India, it played a similar role to the development of a nuclear weapons capacity (examined in a recent thesis by Ulrika Möller). Developing countries have seen nuclear technology as a path to strength and progress. It has not escaped their attention that leaders on the world stage often have such capacity.[29] Furthermore, countries that need more energy to expand their manufacturing sectors and create private wealth may increasingly regard nuclear energy as a matter of equity. Among such countries are Argentina, Brazil, China, Egypt, India, Iran, Pakistan, Romania, Russia, Ukraine and Vietnam. They all see the use of nuclear technology as a means of catching up with North America and Western Europe.

Equity, democracy and technology have all been closely related to national interest, development, politics and economics during the post-war period. Nations are the framework within which demands for justice have been met, democratic institutions have been defended and technologies have emerged. As has been widely discussed in recent years, many of the challenges faced by contemporary society – poverty, war, armament, climate change and environmental threats – require transnational solutions. Though still fundamental social values, equity, democracy and technology now appear in new contexts.[30] Nuclear power companies frequently operate in other countries. For example, the German company E.ON (partly owned by Westinghouse in the United States) plays an important role in the Swedish nuclear industry. The nuclear power plant to replace Ignalina will be built as a collaborative effort of the Baltic countries. India, which has defined national self-sufficiency as a fundamental objective of nuclear energy and waste management, intends to begin privatising the industry. The existence of powerful transnational interests within the industry makes it more difficult to publicly plan the management of nuclear waste, interferes with the emergence of common standards and complicates the issue of responsibility. Finally, technological progress in the management of nuclear waste is already global in scope. Research teams are made up of members from different parts of the world, while businesses and public authorities in various countries work closely together.

(vi) Best available science and technology

Peaceful and commercial use of nuclear power was developed on the assumption that it would be the best available source of energy. Deep geological disposal subsequently appeared to be the most logical approach for managing waste. West Germany concluded that rock salt was the optimal solution and focused on Gorleben as of the 1970s.[31] The Swedish effort, which began in the early 1980s, attracted international attention. Based on the idea of burying waste in bedrock and encapsulating it such that it is surrounded by multiple barriers, the KBS-3 method has been taken over by Finland. The model has also been adopted by the British nuclear waste management programme and has been described in a European Commission report as a full-fledged solution.[32]

Identification of deep geological disposal as the optimal solution has reduced the perceived importance of finding the best possible bedrock. The United States strove for the most scientifically defensible approach but stumbled when choosing a suitable repository site. Barry Solomon reports that Yucca Mountain was selected in the face of strong scientific reservations about its geological stability.[33]

The countries that invested in reprocessing technology needed to manage the subsequent waste. The UK also worked on a deep geological disposal solution in the 1970s, clearly oriented towards finding the best scientific approach. The idea was to first locate the optimum site and then persuade the community involved. Due to local protests against exploratory drilling in 1981, the development of repositories for high-level nuclear waste was shelved for 50 years.[34] The French nuclear waste management programme has also made many earnest assurances that science and technology would generate the best possible solutions. Plans in the 1960s were to deposit waste under the Mediterranean. Deep geological disposal was identified as the optimal solution in the early 1980s.[35]

How does the issue of optimal technology complicate nuclear waste management today? Although current decisions may be well founded and wise, new technologies in the not-too-distant future may easily render them obsolete and faulty. A deep geological disposal programme requires major investments and a multi-year construction effort. Such a facility could be up and running by 2020. But a better technology could conceivably be available by 2030. Another possibility is that demand in the 2020s for spent nuclear fuel for reprocessing purposes will make it more profitable for producers to sell than bury it – even more so if transmutation technology succeeds in reducing or eliminating the plutonium that currently is a by-product of reprocessing. French sociologist Yannick Barthe argues that the focus on deep geological disposal stymies development of other solutions. In recent years, historians of economics and technology have referred to the phenomenon that arises as path dependence. This term describes the tendency of social development to go along with previous approaches and deflect attempts to redirect them. That which is perceived as the best available technological solution locks society into a particular concept.[36]

The concept of risk society can be related to nuclear waste management from this point of view. The deep geological disposal model is consistent with the 'out of sight, out of mind' attitude of industrial society, the only difference being that the mechanism consists of tunnels rather than smokestacks. But the model can also be interpreted as a sensible risk management strategy – countries like the United States and Sweden rejected the reprocessing method because of the risks it entailed. The high awareness of risk that currently prevails has other implications as well. More complete information about the hazards of nuclear waste spurs demands to minimise the risk and ultimately to bury it deeper, in more suitable bedrock, and so on. A similar dynamic may lead to cries for a safe reprocessing method in countries like France and Japan, as well as transmutation plants. Anyone who wants to promote the legitimacy of nuclear waste management should expect growing insistence that the risks be minimised by means of the most suitable science and technology.

The choice of Forsmark as a repository site by the Swedish nuclear waste management programme in 2009 argued that the bedrock was more stable there than in Oskarshamn, the alternative. However, few potential sites were actually explored. The geological standard has been redefined from optimality to adequacy.

This case can be further discussed. Integral to the Swedish geologists' defence of deep geological disposal is that parts of Sweden, like Finland, lie on the Baltic Shield. The bedrock has been around for 1–2 billion years, a long time even in geological terms. Its activity is negligible, very little has happened there for millions of years, and Swedish geologists assume that it will remain stable. 'Thank God for the Baltic Shield', one of them once exclaimed.[37] The first complication is locating the site that is geologically most suitable for disposal purposes. Geologists at the Swedish Nuclear Fuel and Waste Management Company (SKB) originally sought the best possible site. But even exploratory drilling ran up against resistance by local residents and authorities. They switched tactics in the early 1990s and began looking for places that offered satisfactory bedrock and were willing to accept a repository. The fact that SKB identified the Baltic cities of Oskarshamn and Östhammar, both of which have nuclear power plants, aroused suspicion that bedrock quality was no longer a prime consideration. Nevertheless, geologists are increasingly comfortable with the idea of suboptimal bedrock.[38] Consistent with the European Commission's approach to a continent-wide system for managing nuclear waste, SKB now argues that the KBS-3 concept can be extended beyond the Baltic Shield and implemented in many other countries.[39]

One scientific objection to the KBS-3 method involves the copper capsules in which the nuclear waste is to be placed. Sceptical that the capsules will remain intact for a hundred thousand years, some corrosion researchers estimate fifty thousand years instead. They have identified problems with the tests and conclusions of both SKB and Posiva Oy, its Finnish counterpart.[40]

However, the idea of deep geological disposal also revives legitimacy as popular will. The bedrock must be suitable as a repository and be located where local communities and officials are willing to go along. In other words, the waste must be composted in a safe and secure manner while satisfying the legitimate claims of democracy and public opinion. Local communities in the countries that produce nuclear power usually reject such proposals. Critics of deep geological disposal see a conflict between safety and democracy. Vetoes by communities located above the kind of bedrock that best meets safety standards lead to the choice of less suitable sites. On the other hand, geologists dismiss the criticism, arguing that available bedrock meets the criteria with room to spare.

(vii) Communication and dialogue

Once decisions were made to build deep geological repositories, the next question was how to implement the projects. Popular protests stood in the way throughout the 1980s. The waste management effort switched strategy in 1990. Starting in Sweden and France, the focus was on providing exhaustive information about what needed to be done, as well as conducting dialogue and consultation sessions with the local population and representatives of community organisations at the exploration sites. In Sweden, the Nuclear Fuel and Waste Management Co sent a letter to all municipalities requesting invitations to explore possible repository sites. A few municipalities that answered in the affirmative were subsequently chosen. When *Agence nationale pour la gestion des déchets radioactifs* (ANDRA) sent a similar request, 30 mayors expressed interest.[41]

Other countries experienced similar developments, if somewhat later. Germany began to discuss the original process in the late 1990s, leading to objections that it was opaque and top-heavy. However, no fresh direction has yet been staked out. As Hocke and Renn point out from the German case, the new orientation towards communication, dialogue and deliberation focuses not only on how to manage nuclear waste but on how to shape the political process. Controversies surrounding statements by experts are to be handled such that various interests and values are weighed by a democratic decision-making mechanism.[42] Since 2003, as reported by Gordon Mackkerron and Frans Berkhout, the British strategy has been to win widespread trust by means of an open dialogue between stakeholders and the general public. The deliberative process comprises 'citizen panels, discussion groups, a national stakeholder forum, nuclear site stakeholder round tables, a web-based programme and a large school project', including critical voices. As in Sweden, the government's advisory body recommends that communities participate on a voluntary basis and have the right to back out if their conditions are not met.[43] Canada conducted a public inquiry in 1996–97 concerning a deep geological disposal solution. A committee travelled around the country to discuss ethical issues, technical assessments and the views of the local populace. Canadian sociologist Durrin Durant concludes that the subsequent report defined the problem as one of society and management rather than technology and safety. In the 2000s, the Canadian nuclear waste management industry tried to learn from those lessons, using public-opinion surveys and focus groups, as well as dialogue sessions and workshops at the community level. The agencies and government ultimately concluded that any decision should be postponed for at least three decades.[44]

The industry can borrow the widely used concept of corporate social responsibility (CSR) to legitimise nuclear waste management. CSR literature links social responsibility to legitimate management of a business through constant communication with customers and employees.[45]

Durrin Durant, as well as Rolf Lidskog and Göran Sundqvist, argue that neither Canada nor Sweden pursued truly deliberative processes, given that stakeholders had a clear objective from the very beginning. Thus, strategic considerations predominated.[46] An important dilemma arises when legitimacy is pursued in such a manner: what happens if local views change and a repository is rejected? Trust must be created by means of a deliberative process that is also responsive to shifting public opinion. But what do you do if construction of a repository is already under way and the community changes its mind?

The problem illustrated by the examples cited in this chapter is that decisions concerning nuclear waste management proceed from narrow meanings of legitimacy that can bypass crucial values, considerations of the optimal site for a repository or the imperative of thoroughly consulting with the local community. Such decisions may be made within the framework of parliamentary democracy though unconnected to the structures, institutions and power relationships on which civic society is built.

Notes

1 C. K. Ausell 'Legitimacy, Sociology of', in *International Encyclopedia of the Social and Behavioral Sciences.*

2 B. Badie 'Legitimacy: Political', in *International Encyclopedia of the Social and Behavioral Sciences.*

3 Thomas Würtenberger 'Legitimität/Legalität' in *Geschichtliche Grundbegriffe*, Stuttgart: Clett Kotta, 1982, p. 691f.

4 Würtenberger op. cit. pp. 695f.

5 Peter Burke *The fabrication of Louis XIV*, London: Yale University Press, 1992.

6 Würtenberg op. cit. pp. 697. See definition of 'Legitimität' in Hans Schulz *Deutsches Fremdwörterbuch*, Berlin 1926, and *Etymologisches Wörterbuch des Deutschen*, Berlin: Akademie Vgl, 1993.

7 Simonde de Sismondi *Études sur les constitutiones des peuples libres*, Bruxelles 1836 (1815).

8 Max Weber *Wirtschaft und Gesellschaft: Grundriss der Verstehende Soziologie*, Tübingen: Mohr (Siebeck), 1972 (1921), p. 549.

9 Weber op. cit. pp. 122–47.

10 Weber op. cit. pp. 18ff.

11 Ulrich Beck *Risk Society: Towards a New Modernity*, London: Sage 1992 (1986).

12 See Chapter 3, section on theories of legitimacy.

13 The research project 'The Political Challenges of Nuclear Waste Management' was financed by a small grant from the Bank of Sweden Tercentenary Foundation in 2005, while 'Current Trends in Nuclear Waste Management' was financed by Formas 2008–10.

14 Urban Strandberg and Mats Andrén (eds) *Journal of Risk Research* 7–8/2009.
15 Gordon MacKerron and Frans Berkhout 'Learning to listen: institutional change and legitimation in UK radioactive waste policy' in *Journal of Risk Research* 7–8/2009, p. 11.
16 Barry Solomon 'High-level Radioactive Waste Management in the U.S.' in *Journal of Risk Research* 7–8/2009, pp. 6, 11.
17 Lena Andersson-Skog 'Från en energi till farligt avfall – kärnkraftsfrågans reglering i det svenska välfärdsbyggandet. En ekonomisk historisk översikt' in Andrén and Strandberg *Kärnavfallets politiska utmaningar*, Hedemora 2005. Mark Elam and Göran Sundqvist 'The Swedish KBS Project: A Last Word in Nuclear Fuel Safety Prepares to Conquer the World?' in *Journal of Risk Research* 7–8/2009, pp. 20ff.
18 *Partial verdict, 22 March 2006, handed down in Vänersborg*, Vänersborg District Court (Environmental Court)
19 Peter Hocke and Ortwin Renn 'Concerned Public and the Paralysis of Decision Making. Nuclear Waste Management Policy in Germany' in *Journal of Risk Research* 7–8/2009, p. 19.
20 Hocke and Renn op. cit. pp. 6ff.
21 Darrin Durant 'Radwaste in Canada: a political-economy of uncertainty' in *Journal of Risk Research* 7–8/2009, p. 8.
22 MacKerron and Berkhout op. cit. p. 11.
23 Evert Vedung 'Det högaktiva avfallets väg till den rikspolitiska dagordningen' in Andrén and Strandberg op. cit.
24 Elam and Sundqvist op. cit.
25 MacKerron and Berkhout op. cit. p. 10.
26 Elam and Sundqvist op. cit.
27 Solomon op. cit. pp. 23ff. Urban Strandberg and Mats Andrén 'Editorial: Nuclear Waste Management in a Globalised World' in *Journal of Risk Research* 7–8/2009, pp. 20ff.
28 Ram Mohan and Veena Aggarwal 'Spent fuel management in India' in *Journal of Risk Research* 7–8/2009, pp. 3, 16.
29 Ulrika Möller *The prospects of security cooperation: a matter of relative gains or recognition? India and nuclear weapons control*, Gothenburg: University of Gothenburg, 2007.
30 For discussions of the challenges posed by globalisation, see Ulrich Beck *The Cosmopolitan Vision*, Cambridge: Polity, 2006 (2000), Jürgen Habermas *The Postnational Constellation*, Cambridge, MA: MIT Press, 2001 (1998), Peter Kemp *Citizen of the World: Cosmopolitan Ideals for the 21st Century*, Lancaster: Prometheus Books, 2010 (2005).
31 Hocke and Renn op. cit. pp. 6ff.
32 Elam and Sundqvist op. cit. pp. 9, 23f.
33 Solomon op. cit. p. 14.
34 MacKerron and Berkhout op. cit. pp. 12f.
35 Yannick Barthe 'Framing nuclear waste as a political issue in France' in *Journal of Risk Research* 7–8/2009, pp. 2–9.
36 Barthe op. cit. p. 11.
37 Jimmy Stigh 'KASAM och den Baltiska skölden' in *Kärnavfall: tillbakablick och framtidsperspektiv i KASAM:s verksamhet*, Stockholm, SOU 2004:120, pp. 31–36. Quote, p. 35.
38 Gustafson op. cit., Stigh op. cit. Mark Elam and Göran Sundqvist 'Meddling in Swedish Success in Nuclear Waste Management' in *Environmental Politics* 2, 2011.
39 Elam and Sundqvist op. cit.
40 Torbjörn Åkermark 'Oseriöst bygga slutförvar för kärnavfallet redan nu', DN.Debatt, 31 July 2010. Digby MacDonald and Samin Sharifi-Asl 'Is Copper Immune to Corrosion When in Contact with Water and Aqueous Solutions?' SSM Report 2011–09.

41 Barthe op. cit. p. 3. Elam and Sundqvist op. cit. pp. 14ff. Barthe op. cit. p. 18f.
42 Hocke and Renn op. cit. pp. 18ff, 26.
43 MacKerron and Berkhout op. cit. pp. 18f.
44 Durant op. cit. pp. 15f, 20ff.
45 Elisabet Garriga and Domènec Mele 'Corporate Social Responsibility Theories: Mapping the Territory' in *Journal of Business Ethics* 53, 2004, pp. 51–71.
46 Durant op. cit., p. 6. Rolf Lidskog and Göran Sundqvist 'On the right track? Technology, geology and society in Swedish nuclear waste management' in *Journal of Risk Research* 7–8/2004.

3 Ethics and legitimacy

Exploring the correlation between legitimacy and ethics helps illuminate the problems associated with managing nuclear waste. Furthermore, ethical arguments are fundamental to analysing responsibility issues. The thesis is that ethical concerns demand an expansion of the concept of legitimacy. This and the following chapter set out a conceptual framework for analysing and discussing ethical issues of nuclear waste management. This chapter first examines the ethical arguments associated with nuclear waste management, and then looks at the general relationship of ethics to legitimacy. The next chapter explores nihilism and responsibility.

Ethical principles in place

Nuclear waste management must meet a series of criteria to increase stable legitimacy over the next few centuries. In addition to the meanings of legitimacy identified by the preceding chapter, the discourse on nuclear waste since the 1980s and 1990s has included ethical arguments as well.

Those who have sought support for various nuclear waste management solutions have regarded ethical arguments as important. The most obvious example is what has been referred to as the ethical principle – the idea that each country and generation should assume responsibility for managing the nuclear waste it produces. The principle has been particularly prominent in some countries in North America and Western Europe, ironically during the period that the nuclear fuel cycle has been internationalised in the wake of privatisation and economic collaboration.

Part of the background of this principle is the new focus on geological disposal after ocean dumping was banned by an international convention in 1975 and the United States abandoned reprocessing of nuclear waste in 1977.[1] Environmental awareness since the 1970s has highlighted ethical considerations, followed by environmental research and the sustainable development debate of the late 1980s.

In Sweden the foundation for such an ethical principle was laid down in the mid-1980s. The idea of national responsibility was launched in parliamentary discussions and proposals,[2] while the Swedish National Council for Nuclear

Waste (formerly KASAM) declared the ethical principle, including the requirement that responsibility be assumed by the present generation. The French took a similar step in the early 1990s, emphasising new ethical principles, including the rights of future generations, as well as protection of health, nature and the environment. The US Congress enunciated several principles in 1997, including protection of health, safety and future generations.[3] In Canada the Nuclear Fuel Waste Act of 2002 required ethical considerations.[4] In Switzerland and the Netherlands, ethical principles have been presented as well.[5]

Documents issued by the International Atomic Energy Agency (IAEA) and Nuclear Energy Agency (NEA) in 1995 established basic ethical principles for nuclear waste management. In addition to safety, environmental and monitoring considerations, the documents addressed the issue of responsibility for future generations. These principles were also accepted by the World Nuclear Association (WNA).[6]

In the IAEA document the principles are presented in the following way:

> Principle 1: Protection of human health
> Radioactive waste shall be managed in such a way as to secure an acceptable level of protection for human health.
> Principle 2: Protection of the environment
> Radioactive waste shall be managed in such a way as to provide an acceptable level of protection of the environment.
> Principle 3: Protection beyond national borders
> Radioactive waste shall be managed in such a way as to assure that possible effects on human health and the environment beyond national borders will be taken into account.
> Principle 4: Protection of future generations
> Radioactive waste shall be managed in such a way that predicted impacts on the health of future generations will not be greater than relevant levels of impact that are acceptable today.
> Principle 5: Burdens on future generations
> Radioactive waste shall be managed in such a way that will not impose undue burdens on future generations.
> Principle 6: National legal framework
> Radioactive waste shall be managed within an appropriate national legal framework including clear allocation of responsibilities and provision for independent regulatory functions.
> Principle 7: Control of radioactive waste generation
> Generation of radioactive waste shall be kept to the minimum practicable.
> Principle 8: Radioactive waste generation and management interdependencies
> Interdependencies among all steps in radioactive waste generation and management shall be appropriately taken into account.
> Principle 9: Safety of facilities
> The safety of facilities for radioactive waste management shall be appropriately assured during their lifetime.[7]

According to both documents, nuclear waste 'shall be managed in such a way that predicted impacts on the health of future generations will not be greater than relevant levels of impact that are acceptable today' and that it 'will not impose undue burdens on future generations.' In other words, the onus of waste management was to be borne by the members of a particular generation who benefitted from nuclear energy production.[8]

The NEA document does not focus on national responsibility. It defines three factors behind the responsibilities to present and future generations. The first is: 'the ethical principles of intergenerational and intragenerational equity', the second: 'the technical requirements to ensure, and give confidence in, safety now and in the future', and the third: 'the availability of resources for technology development and implementation'.[9]

After 1995, references to ethics grew common as a basis of legitimacy for nuclear waste management. In the era of globalisation, the principle of national responsibility from which current management of spent nuclear fuel proceeds is rather shaky. National responsibility is difficult to uphold when fuel is traded on the global market and international companies have significant holdings in nuclear reactors; the Treaty Establishing the European Atomic Energy Community (Euratom) states the European Union's responsibility for nuclear fuel assets. The principle of generational responsibility tends to be unrealistic as well. The notion that each generation must manage its own nuclear waste has quickly become inoperable now that industrialised countries have generated waste for 50 years and simply stored it in repositories for future societies to deal with. The public officials who decided to continue producing nuclear energy after the 1980 referendum in Sweden would have to be 150 years old before the planned repository was completed.

The previous chapter mentioned Ulrich Beck's distinction between the grounds on which legitimacy is based in industrial society and what he refers to as risk society. He argues that the kind of rationality inherent to risk society is more inclined to raise fundamental ethical questions. His concept that the rationalities of the two societies exist side by side sheds light on ethical issues and challenges the technological thinking of industrial society.[10]

Analogously, legitimacy for nuclear waste management can be viewed as either a scientific/technological or an ethical question. Geologists are satisfied if they can find bedrock that is suitable to bury spent nuclear fuel for 100,000 years. According to their models, that time frame ensures adequate safety. From an ethical point of view, the more important consideration is that waste be stored under the safest possible conditions. Such an approach dictates the search for the most stable bedrock. This is one reason to question the seriousness of nuclear waste management programmes when it comes to their interest in ethical principles.

The very impact of such ethical principles can be discussed. Hocke and Renn argue that they were of less importance in Germany. Occasionally it is said that the present generation should solve its own problems, but the principles have not become a major point of contention. Protest organisations have raised the equity issue in terms of allocating benefits and costs between generations, but it has long been neglected in Germany by official organisations, industry and parties.[11] On the other hand, the Swedish National Council for Nuclear Waste considered the ethical principle and demanded additional ethical dimensions at a fairly early stage. Their responses to government reports, as well as reports on the nuclear waste programme of the Swedish Nuclear Fuel and Waste Management Company (SKB), are inclined to raise the issues of accountability and national responsibility.[12]

It is clear that ethical arguments were presented in several countries in order to contribute to the creation of legitimacy for nuclear waste management, but this does not mean that ethical concerns had an impact on the considerations behind the programmes of the various countries.

One should question the outcome of the ethical considerations of the parties mentioned here. For Sweden, it is said to be geological disposal. Being in favour of deep geological disposal, the Swedish Nuclear Fuel and Waste Management Co/SKB rejects the transmutation alternative with reference to 'the requirement of the IAEA's radioactive waste convention, i.e. that undue burdens shall not be imposed on future generations', as it will take too long to develop the option.[13] The Swedish National Council for Nuclear Waste argues that deep geological disposal is the possible option that gives the best correspondence when ethical demands are considered.[14] The IAEA and NEA come to similar conclusions as well. The NEA says 'geological disposal of long-lived radioactive wastes takes intergenerational equity issues into account, notably by applying the same standards of risk in the far future as it does to the present, and by limiting the liabilities bequeathed to future generations'.[15] It is hard to avoid the conclusion that ethical arguments are used to defend the solution of deep geological disposal.

Obviously, the ethical principles are biased towards geological disposal. They seem to represent an investment aimed at selling a solution, to paraphrase a report by Digby C. Andersson, or at least sell the image of a possible solution.[16] A report by Malcolm Grimston for the Committee on Radioactive Waste Management/CORWM concluded that the introduction and use of ethical arguments have not changed policy; rather they have provided support for the consultation process.[17]

There are good reasons for arguing that the solution of geological disposal is not ethically sound. Kristin Shrader-Frechette has made a strong ethical critique of the geological disposal project at Yucca Mountain. As already described in Chapter 1, she argues that deep geological disposal entails a utilitarian way of thinking, underestimating the risks and uncertainties for future generations.

In our discussion of the nuclear waste issue, we will show that the ethical analysis by proponents of long-term geological disposal is wrong. Shrader-Frechette similarly distrust the nuclear waste programme of the United States. But she does not outline a conceptual framework of legitimacy and responsibility for nuclear waste management, which we will do before embarking on a discussion of the deep geological solution.

Why legitimacy and not trust

During the period that nuclear waste management has been commonly linked to ethics, the concept of trust has assumed prominence among the methods used to study the issue, as well as the programmatic recommendations presented by researchers and industry representatives. Discussion of trust has thereby served as a means of promoting and accelerating implementation of the deep geological disposal solution. A 1999 comparative study of the United States, Britain, France and Germany by Andrew Blowers concluded that any workable solution must be subject to acceptance by the citizenry and stringent technical requirements.[18] Ever since that time, a large body of researchers have examined the various ingredients of trust, as well as the degree of trust that people have in different approaches to nuclear waste management, as a basis for their recommendations.[19] Blowers is certainly right about this, but one should also remember that research on nuclear waste management can most often be described as intertwined with the activities of stakeholders.

The most serious problem is that the research themes and aims are identical with those formulated by industry consultants and representatives. The same pattern may be observed when it comes to the idea of deliberations and public participation in the process. The NEA, which has advocated civic involvement since the early 1990s, arranged five meetings from 1991 to 1997 for that purpose.[20] A similar kind of unanimity has often characterised the discussion of ethical principles, which have been cited off and on since the 1970s. Not until the IAEA and the NEA started talking about ethics have they become paradigmatic, though frequently referred to in very general terms with the objective of creating consensus and legitimacy for deep geological disposal.[21]

All this has to do with the 'proximity between social scientists' research approaches and the policy standards of the focus area', as argued by Solomon, Andrén and Strandberg, which is a distinctive feature of this research area. As nuclear waste management research was originally a technical subject, it started as technical projects and used 'technical knowledge and engineering ideals of finding practical solutions such as deep geological repositories. It was only when real-world hindrances of a social character occurred that the usage of social science research came into sharp force'.[22]

There is a need for research on the nuclear waste issue that is conducted at arms length from the industry.

When a cry for ethics is combined with the assumption that general distrust has prevented implementation of the deep geological disposal solution, a reasonable suspicion is that the motivation is to generate such trust based on a consensus about principles that are shared and unshakable. The next step is to describe the deep geological disposal solution in a manner that corresponds to such principles. As a tool for building consensus, ethics are incorporated into the solution itself.

Why is legitimacy a better concept than trust? And why does the concept of trust narrow the prospects for using ethics in a critical way? Why is it important to regard nuclear waste management in the light of both legitimacy and ethics?

Making Democracy Work: Civic Traditions in Modern Italy (1993) and *Bowling Alone: The Collapse and Revival of American Community* (2000) by Robert Putnam brought the concept of trust to the fore in current political science. The books argue that trust in community institutions is fundamental to a well-functioning society. Putnam emphasises social capital above all. The first book concludes that the new regional governments established in 1970 have generally been much more effective in northern than southern Italy due to greater trust in community institutions. Quite simply, citizens must be able to rely on the presumption that the market, judicial system and other components of the society they live in are working properly. According to Putnam, northern Italy has a long tradition that stretches back to the Renaissance and stresses both civic involvement – particularly political participation – and institutions to channel it.[23] Putnam has thereby made a major contribution to the tendency among political scientists to examine the mechanisms of trust and how it is generated.

In view of the previous chapter, trust may be regarded as a reformulation of the legitimacy concept. Trust stresses the role of tradition, civic responsibility, a strong legal framework and dialogue between citizens and public officials in a well-functioning democracy. However, issues surrounding technological imperatives and their conflicts with democratic governance reveal the limitations of the trust concept. For instance, it may be argued that the use of nuclear technology in health care has saved more lives than were destroyed in Hiroshima and Nagasaki. The limitations of trust arise when considering the negative consequences that cannot be foreseen when a technology, which may not have been subject to democratic decision-making processes, is commercialised. Even technology that is implemented in a democratic manner affects people who live in a different geographic region or belong to a future generation.

Nuclear technology and waste are among the issues that require a broader concept than trust. Legitimacy is a more fruitful concept than trust in that it is open to the critical dimensions of ethics. The implications of that assertion will be examined shortly. The crucial point here is the greater suitability of the legitimacy concept when exploring ethical matters.

There are additional reasons for being suspicious of – if not downright opposed to – the use of trust in this context. The concept gained currency in the management and marketing literature of the early 1990s: 'Trust is commonly regarded as crucial since its roles or functions for the wellbeing of business relationships are cardinal'.[24] Thus, the concept is a well-developed approach that is highly suitable for industry representatives who want to promote nuclear waste management projects.

Just because a concept appears in non-scientific literature does not mean that it is irrelevant. On the contrary, the use of trust in management and marketing research may hold important lessons for political scientists.

A more revealing approach is to examine how loyalty relates to and serves as a basis for legitimacy. Loyalty could be an obsolete way of looking at legitimacy. Medieval and feudal society was based on a system of loyalty between kings, lords and vassals. Relationships (guilds and workshops) in the pre-modern communities that preceded civic society were built around loyalty as well. However, the concept is also useful for the understanding of legitimacy in modern multi-level society, which makes decisions and creates identities through a variety of tightly interwoven communities. National citizenship is certainly a strong foundation on which to construct legitimacy. But transnational communities, of which the EU is the paramount example, programmatically strive to expand their legitimacy. Furthermore, the cultures that are rooted in local communities are fertile grounds for creating legitimacy. Finally there's the world community, along with the growing realisation and need to act jointly. The idea of global community implies a loyalty to the human race that transcends national or local interests. Thus, an in-depth analysis of loyalty as a foundation for legitimacy would be highly useful. A loyalty-based concept of legitimacy would have to consider the various types of communities and their interrelationships.

Loyalty may also be linked to future generations – in other words, it need not be synchronous, but may also be diachronous and forward-looking. Thus, loyalty is more compatible with ethics as a basis for legitimacy than trust is. This compatibility is particularly obvious when addressing issues of responsibility. Trust is not associated with responsibility. Someone who trusts a political regime feels no responsibility for it – usually just the opposite. Trust is based on the assumption that the regime will live up to its commitments vis-à-vis the citizenry. However, loyalty can readily be linked to responsibility for the wellbeing of both current and future generations.

Kant and the dualism of legality and morality

Discussing ethics and legitimacy together is nothing new. The early use of the legitimacy concept reveals a close relationship between the two. A concept of legitimacy from Kantian philosophy bases governance on ethics. That is where legitimacy and ethics are wedded.

Kant published his last great work *The Metaphysics of Morals* in two volumes (1797–98) in the wake of the upheavals in France. Without using the actual word, he brushes against the modern concept of legitimacy when he declares that 'the legislating authority can belong only to the united will of the people'.[25] He links that will to citizens and assigns them three characteristics: the freedom to pass laws, equality, and independence from each other. The third characteristic is the basis on which Kant distinguishes between active and passive citizens. All citizens are entitled to have their voices heard, but only those who are independent should actively participate in the state's business, organise its institutions and be involved in the legislative process. The citizens he imagines are members of the bourgeoisie whose social position is based on private property. Workers, journeymen, servants, women and children are out of the question. Thus, he manages to incorporate the modern concept of legitimacy while adapting his thought to conditions in post-revolutionary Europe.[26] From the perspective of more than two centuries, Kant was clearly proceeding from middle-class citizens of the late 1700s. Thus, his attempt to base his philosophy of law and notion of obedience on the universal principles of pure reason was a product of his time, with all the limitations that entails. But much of his philosophy is forward-looking as well.

Although Kant does not speak directly about legitimacy, he uses legality – the older word. His combination of legality and ethics, of basic importance for legitimacy, was a new approach that later had a major impact on legal philosophy. Kant places all legal and judicial matters within the context of a society's explicit laws. They are among the external constraints to which citizens are bound without needing to understand them. But citizens may also take internal, ethically based laws into consideration. Kant does not assume that legal and ethical principles necessarily diverge, but stresses the conflict that they can engender. Legal and ethical behaviour may be based on different kinds of obligations and concepts. Ethical obligations always have an internal basis, whereas legal obligations inevitably have an external basis in addition to a possible internal basis. Adhering to a contract is an external obligation if it was drawn up in accordance with the law and an internal obligation if external motives are disregarded. Ethical action proceeds from an internal obligation, whose importance Kant emphasises by characterising it as a type of law distinct from that imposed by society or nature.[27] This internal obligation includes loyalty towards certain principles or values.

Kant does not make it crystal clear whether legality or ethics should be given priority. Some evidence suggests that he puts ethics first. For instance, he says that everyone should strive for self-enlightenment and trust their own reason. But more recent research has suggested that his political philosophy tends to stress legality. He presents his famous concept of peaceful coexistence between nations as a legal rather than an ethical principle. The research highlights the pluralistic aspect of his thought: the idea that practical reason does not possess final solutions, but arrives at wise decisions by confronting the views and perspectives of other people.[28]

Post-Kantian German idealistic philosophy made the distinction between external and internal obligation in ever-changing ways. Fichte posits a dichotomy between legal and ethical principles. Similarly, Hegel assigns legal principles to the state and ethical principles to the religious community. But he combines them in an 'ethical community', the customs and institutions that arise from the temperament and character of a society. For him, legality must coincide with objective reality in order to be regarded as legitimate.[29]

The concept of legitimacy that stemmed from Kantian and German idealistic philosophy stressed the importance of basing governance on an ethical principle or community (morality). German idealistic philosophy had to link legality and morality in order to establish legitimacy. In other words, that correlation has been around for a long time.

The concept of ethics harbours a dilemma. On the one hand, it must provide a stable component around which long-term legitimacy can be constructed. On the other hand, it must have a foundation of its own. The foundation for Kant was the universal reason that he analysed from various points of view, but the bourgeois individual was always his point of reference. His reason-based republic of citizens proceeded from the will of the educated bourgeoisie rather than that of the people. This is where his ethical component resides. In Hegel's case, the foundation of ethics is society, along with the institutions that emerged throughout history as deciphered by philosophers. One doesn't have to go far to find objections to both points of view. Nobody has ever been able to offer ethical guidelines that gain universal acceptance. Normative systems have conflicted with each other throughout history. No single thinker can achieve ethical supremacy. Societies have always been complex when it comes to both politics and basic identity. The collision of normative systems has accelerated in the age of globalisation, reinforcing the need for common, international principles.

Ethics and theories of legitimacy

What is the foundation of legitimacy in today's world? While legitimacy based on the nation-state is a legacy of the twentieth century, the twenty-first century is characterised by globalisation. Politics, economics and law all have strong national features, which are merging with transnational trends. This is not the place to discuss whether legitimacy arising from the nation-state is a blind alley that globalisation will eventually render obsolete or whether it remains relevant. The question that concerns us here is whether the current ethical discourse implies a new concept of legitimacy.

Within social theory, positivistic approaches have avoided judgemental and normative definitions of legitimacy. However, the most recent theories consider ethics once again.

The concepts of legitimacy that derive from various traditions of jurisprudence are a good place to start. The relationship of legitimacy to legality is of paramount importance in this respect. One positivistic tradition has regarded the two concepts as identical. The law is an isolated system that requires no support from politics or any external set of principles. At the other end of the spectrum is an instrumental approach according to which law reflects the political system and nothing else. Legitimacy is reduced to political manoeuvring and the values it represents. Hans Kelsen is the most prominent spokesman for the first perspective and Carl Schmitt for the second. The discussion of legitimacy and legality in recent decades has also drawn nourishment from a liberal critique of legal positivism. Johan Rawls stresses the role of liberal values in the establishment of legal legitimacy, while Habermas argues that law creates legitimacy in dialogue with other areas of society.[30]

An influential theory presented by political scientist David Beetham in 1991 proceeds from three dimensions of legitimacy. While power must be exercised in accordance with legality, it must also be based on widespread perception of the political order as responsive to the public interest (normative justifiability) and be generally accepted (legitimation). The model links legitimacy to power in the broadest sense, including both the political and social spheres. But legitimacy cannot be secured from general ethical principles, other than those that are directly related to the political order. Beetham emphasises the importance of underlying normative principles for constitutionalism and similar phenomena but does not examine other principles that laws and policy-making must consider (unless included in the concept of general acceptance, which would comprise several meanings of legitimacy discussed in the previous chapter).[31] Because Beetham's theory specifies the demands that a social order must meet in order to gain legitimacy but says little about the performative aspects of legitimacy, it does not contain an ethical argument. The model focuses on the aspects of legitimacy that are historically stable – its reflection of power relationships in political and social life and its foundation in common norms for the institutional framework of the political order. But it says nothing about how legitimacy is forged.

When Beetham departs from the vagueness of his general model and looks at concrete societies (liberal democracies), he can also observe how legitimacy is forged in relation to historical circumstances. Thus, normative justifiability has three pillars: the creation of identity that includes human beings, a democratic order that bestows authority on elected officials, and satisfaction of the needs of citizens. Beetham's analysis of the ways that the EU creates legitimacy identifies clear attempts to generate a European identity and expand the authority of the European Parliament, while providing security, general welfare and civil rights.[32] Legitimacy in this sense can be conceived of as a situational concept that accumulates meanings as it goes along.

Habermas's examination of legitimacy also stresses the importance of fundamental consensus about the political order. Legitimacy is based on norms that emerge on an ongoing basis rather than being established once and for all. In one important sense, his association of legality and morality transcends the limitation of legitimacy theory to the political order that is typical among political scientists. Law and politics are no longer viewed as hermetically sealed and self-referential systems. On the contrary, their rules and norms grow out of a communicative sphere or discourse that must speak to a broader public in order to work. Although law and politics are institutionalised systems, they reside in 'a network of communicative actions'. To understand how the norms of modern society develop within specialised spheres such as law and politics, the communicative conditions that link them to the life-world must be considered. Conversely, Habermas argues that government policies that are not grounded in democratic institutions and public consent will encounter legitimacy problems.[33]

Habermas's thesis is that legality can create legitimacy only if it is anchored in a moral discourse that goes beyond the authority and principles inherent to a particular political order.[34] Many other far-reaching questions may elicit moral standpoints that achieve validity only when everyone is in agreement.[35] One such question might be the role and consequences of technology when it comes to nuclear waste management, which would permit legitimacy and ethics to be linked. When ethics is understood as equity in the Kantian tradition, its place in Habermas's concept of legitimacy is even more striking. He sees ethics as hidden in the discourse and discerns ideas of equity and solidarity in the very terms of the discourse.[36]

Sara Stendahl, in a recently presented thesis in law, describes the mechanism by which politics, morality and administration of the law join together to create legitimacy in the welfare state, providing a means of making Habermas's theory more concrete. She says that there are two sides to the law: formal administration in the courts and general acceptance of its role in upholding justice.[37]

Social theories look at the functional aspects of legitimacy. Of central importance is a fundamental consensus that the basic political order is reasonable and capable of ensuring a good society (Beetham and Habermas). The functional perspective can be fully reconciled with the various historical meanings of legitimacy by seeing them as a specific example of the general case and by asking in the spirit of Beetham how normative justifiability is achieved. Thus, the various functions of legitimacy must meet the expectations or demands that the meanings suggest. The specific example here is nuclear waste management. Thus, normative justifiability will have three cornerstones. The first cornerstone is that legitimate nuclear waste management be consistent with the principles of democratic society. Legality and popular will, the first two meanings of legitimacy, come in at this point. The second cornerstone is that nuclear waste be managed in an inclusive manner such that the citizens concerned have the opportunity to take part in planning and implementation. Inclusion corresponds to the meanings of emotion, tradition and communication/dialogue/deliberation. The third cornerstone is that waste be managed responsibly, which relates to the meanings of emotion, values, optimal science and technology, and ethics. Such ethical responsibility may transcend and extend across the borders of individual societies and generations into the distant future. This is the point at which ethical concerns demand an expansion of the legitimacy argument.

Notes

1 Barry D. Solomon, Mats Andrén, and Urban Strandberg, 'Three Decades of Social Science Research on High-Level Nuclear Waste: Achievements and Future Challenges,' *Risk, Hazards & Crisis in Public Policy*, 1:4, 2010.

2 Per Cramér, Thomas Erhag and Sara Stendahl *Nationellt ansvar för använt kärnbränsle*, Stockholm, Santérus Academic Press, 2009, pp. 19–34.

3 Yannick Barthe *Journal of Risk Research* 7–8/2009 p. 942. Barry Solomon *Journal of Risk Research* 7–8/2009 pp. 1015f.
4 Malcolm Grimston 'Ethical and environmental principles', Committee on Radioactive Waste Management CORWM 2004, pp. 16f.
5 Grimston op. cit. pp. 26, 29f.
6 IAEA 'The Principle of Radioactive Waste Management', Vienna 1995. NEA 'The Environmental and Ethical Basis of Geological Disposal of Long-Lived Radioactive Waste', Paris, 1995. WNA 'Environmental and Ethical Aspects' see http://www.world-nuclear.org/info/Environmental_Ethical_Aspects_inf04ap5.html.
7 IAEA op. cit.
8 NEA op. cit.
9 NEA op. cit. In the NEA document the principles are presented in the following way:

Consideration of these concerns leads to a set of principles to be used as a guide in making ethical choices about waste management strategy:

- the liabilities of waste management should be considered when undertaking new projects;
- those who generate the wastes should take responsibility, and provide the resources, for the management of these materials in a way which will not impose undue burdens on future generations;
- wastes should be managed in a way that secures an acceptable level of protection for human health and the environment, and affords to future generations at least the level of safety which is acceptable today; there seems to be no ethical basis for discounting future health and environmental damage risks;
- a waste management strategy should not be based on a presumption of a stable societal structure for the indefinite future, nor of technological advance; rather it should aim at bequeathing a passively safe situation which places no reliance on active institutional controls.

10 Ulrich Beck *The Cosmopolitan Vision* Cambridge: Polity, 2006 (2000).
11 Peter Hocke and Ortwin Renn 'Concerned Public and the Paralysis of Decision Making. Nuclear Waste Management Policy in Germany' in *Journal of Risk Research* 7–8/2009, p. 52.
12 For reports and reviews of the council, see http://www.karnavfallsradet.se/publikationer.
13 Swedish Nuclear Fuel and Waste Management Co 'RD & D Programme 2007: Programme for research, development and demonstration of methods for the management and disposal of nuclear waste', Stockholm, 2007.
14 Swedish National Council for Nuclear Waste 'Kunskapsläget på kärnavfallsområdet 2007: nu levandes ansvar, framtida generationers frihet', Stockholm, SOU 2007:38.
15 NEA op. cit.
16 Digby C. Andersson *What Has Ethical Investment to Do with Ethics*, 1998.
17 Malcolm Grimston op. cit. p. 15. Compare Mark Elam and Göran Sundqvist 'The Swedish KBS Project: A Last Word in Nuclear Fuel Safety Prepares to Conquer the World?' in *Journal of Risk Research* 7–8/2009.
18 Andrew Blowers 'Nuclear waste and landscapes of risk', see *Landscape Research* 3, 1999.
19 See Barry D. Solomon, Mats Andrén, and Urban Strandberg, 'Three Decades of Social Science Research on High-Level Nuclear Waste: Achievements and Future Challenges', *Risk, Hazards & Crisis in Public Policy*, 1:4, 2010 for an overview of the research. See for example Gilbert W. Basset, Hank C. Jenkins-Smith and Carol Silva 'On site storage of High-Level Nuclear Waste: Attitudes and Perceptions of Local Residents' in *Risk Analysis* 3, 1996, pp. 309–19; Kevin R. Ballard and Richard G. Kuhn 'Developing and Testing a Facility Location Model for Canadian Nuclear Fuel Waste' in *Risk Analysis* 6, 1996, pp. 821–32.

20 H. Riotte 'Stakeholder Issue at OECD/NEA' in JHPS/NEA Symposium, Stakeholder Involvement on Radiation Protection', 2005. See http://www.soc.nii.ac.jp/jhps/s/events/kikaku/stake-sympo/1-Riotte.pdf.
21 IAEA op. cit., NEA op. cit.
22 Solomon *et al.* op. cit.
23 Robert Putnam *Making Democracy Work. Civic Traditions in Modern Italy* (1993) Princeton: Princeton University Press.
24 Lars Huemer, George von Krigh and Johan Roos 'Knowledge and the concept of trust' in *Knowing in Firms: Understanding, Managing and Measuring Knowledge* eds Georg von Krogh, Johan Roos and Dirk Kleine, Sage: London 1998.
25 Immanuel Kant *Die Metaphysik der Sitten*, Frankfurt am Main: Suhrkamp Verlag, 1956 and 1977 (1797), pp. 432: 'Die gesetzgebende Gewalt kann nur dem vereinigten Willen des Volkes zukommen.'
26 Kant op. cit. pp. 432ff.
27 Kant op. cit. pp. 324ff.
28 Carola Häntsch 'The World Citizen from the Perspective of Alien Reason: Notes on Kant's Category of the *Weltbürger* according to Josef Simon' in Rebecka Lettevall and My Klockar Linder (eds) *The Idea of Kosmopolis: History, philosophy and politics of world citizenship*, Huddinge: Södertörn Högskola 2008.
29 Compare Thomas Würtenburger 'Legitimität, Legalität' in *Geschichtliche Grundbegriffe*, Stuttgart: Clett Kotta 1982, pp. 712ff.
30 Sara Stendahl *Communicating Justice Providing Legitimacy: the legal practices of Swedish administrative courts in cases regarding sickness case benefit*, Uppsala: Iustis förlag, 2003, pp. 66f.
31 David Beetham *The Legitimation of Power*, Basingstoke: MacMillan 1991, pp. 16–25, 39ff, 126.
32 David Beetham and Christopher Lord *Legitimacy and the European Union*, London: Longman, 1998.
33 Jürgen Habermas *Faktizität und Geltung: Beiträge zur Diskurstheorie des Rechts und des demokratischen Rechtsstaats*, Frankfurt am Main: Suhrkamp Verlag, 1992, pp. 427ff, 541f, 563ff. Quoted from p. 429. Compare with p. 427: 'Das rechtstaatlich verfasste politische System ist intern in Bereiche administrativer und kommunikativer Macht differenziert und bleibt zur Lebenswelt hin geöffnet. Denn die institutionalisierte Meinungs-und Willensbildung ist auf Zufuhren aus den informellen Kommunikationszusammanhängen ser Öffentlichkeit, des Assoziationswesens und der privatsphäre angewiesen. Mit anderen Worten, das politische Handlungssystem ist in lebensweltliche Kontexte eingebettet.'
34 Habermas op. cit. p. 542.
35 Jürgen Habermas *Den moraliska synpunkten*, 2008 Göteborg: Daidalos, p. 119.
36 Habermas op. cit. pp. 119ff.
37 Stendahl op. cit. pp. 62ff, 77.

4 Nihilism and responsibility

Like legitimacy, concepts of nihilism and responsibility go further back than the issue of nuclear waste management. Nevertheless, we will examine the normative aspects of the issue on the basis of those concepts. Is the production of nuclear waste responsible? Is it nihilistic? The concepts can also shine light on the main solution that has been proposed. Is deep geological disposal consistent with the notion of responsibility? This chapter commences a reconstruction of the responsibility concept. It looks at nihilism and responsibility as historical concepts, in order to set the stage for the next chapter's normative discussion of responsibility and its implications for nuclear waste management. A straight line runs from cultural critics in the interwar years, as well as the 1940s to 1960s when nuclear technology made its breakthrough, to the idea of a new kind of responsibility that emerged as the consequences of nuclear technology became evident.

God is dead

In *The Brothers Karamazov* (1880), Smerdyakov famously says that 'if there's no everlasting God, there's no such thing as virtue, and there's no need of it', claiming that he learned the precept from Ivan. A voice in a feverish dream confronts Ivan with the consequences of his ideas:

> The new man may well become the man-god, even if he is the only one in the whole world, and promoted to his new position, he may light-heartedly overstep all the barriers of the old morality of the old slave-man, if necessary. There is no law for God. Where God stands, the place is holy. Where I stand will be at once the foremost place … 'all things are lawful' and that's the end of it![1]

The dream is one of the many passages in Dostoevsky's works that portray the elements of isolation, solitude, godlessness and lack of moral compass confusion that European civilisation brings to modern life. All that remains is nihilism. Thus, we define nihilism as a lack of loyalty towards ethical principles.

The nihilistic strand in the nuclear waste issue

Nuclear technology may be fodder for the wildest dreams, but it is also a source of nightmares. The alleged role of nuclear technology in bringing the Second World War to a close has long been a matter of controversy. Instead of lasting peace, Hiroshima and Nagasaki were followed by an unprecedented race to stockpile nuclear weapons. The zeal of peace movements and disarmament negotiators reflected widespread awareness of their capacity for destruction on a scale the world had never known.

Nuclear weapons can cause immense suffering and end life on earth as we know it. The world has been infected with a kind of nuclear nihilism ever since the end of the Second World War. The unprecedented arms race has left a number of countries with the capacity to annihilate each other many times over and the danger of regional nuclear conflicts has grown. The strategy of mutual deterrence succeeded during the Cold War. In that sense, the politicians who approved nuclear armament exhibited a kind of responsibility. Given the possibility of human error, however, the risk of total annihilation remains substantial. The very production of nuclear weapons implies a considerable measure of nihilism.

Nuclear energy production also has nihilistic features, though mixed with elements of responsibility. Nuclear power gives rise to nightmares that radiation leakage, though imperceptible to the senses, will damage life and the environment. Reactors have leaked radioactivity, operated beyond their useful life (such as Ignalina in Lithuania), suffered major accidents, or experienced meltdowns. The empirical risk for major meltdowns after Fukushima is roughly 1 per cent of the civilian and commercial reactors that have been built. Translated into figures, it includes one reactor at Three Mile Island, one at Chernobyl and three at Fukushima that suffered explosions and serious damage, out of 440 reactors in operation worldwide and 129 shut down.[2] Storage facilities have not always been in satisfactory condition, and the storage pool at Fukushima was close to emitting large quantities of radiation after the 2011 tsunami in Japan. A fear is that accidents on a par with Chernobyl in 1986 and Fukushima in 2011 are inevitable and will leave their mark on succeeding generations. Risks and accidents appear to be inherent to the nuclear generation. There have been many major accidents: 99 since 1952 and 57 between Three Mile Island in 1979 and Fukushima in 2011, as counted by Benjamin Sovacool using US government definitions.[3] Some analysts warn that the risks are increasing as reactors remain operational too long and bigger reactors with new designs are being introduced.[4]

Meanwhile, the IAEA heads a global collaborative effort to ensure safe production. Producers have taken responsibility for isolating waste from the biosphere. But a big proviso, which has been described as the Achilles heel of nuclear energy production, is in order.

While Sweden and Finland are hopeful about deep geological disposal in tandem with the KBS-3 encapsulation method, no country has yet adopted a workable, long-term solution to the problem of nuclear waste. An overview of the current situation suggests that countries are either groping for solutions or doing their best to avoid them.[5]

The absence of a solution to the problem of nuclear waste after decades of production can be related to the concept of nihilism in a number of ways. The most obvious and serious way is that nuclear power has been developed and nuclear waste has been generated uninterruptedly for more than half a century in the absence of a workable solution or a good plan for its disposal. This is not the place to review the history of nuclear energy and waste, but apparently the dream of a powerful energy source has spawned a nightmare of its own.

Society has irresponsibly averted its eyes from the growing tanks and barrels of waste. The focus on instant gratification, not to mention short-term economic thinking and a worldview that lacks empathy with the plight of future generations, has driven demand in the West. However, some arguments point in a different direction. Nuclear energy has contributed to the prosperity, even among disadvantaged members of the populace, that industrialised countries have created in the post-war period. Most countries that have developed nuclear energy are also welfare states. Furthermore, it can be argued that nuclear waste does not contribute to the greenhouse effect the way that oil and coal do. We will return to the issue of responsible nuclear waste management.

Nihilism and the critique of civilisation

Nuclear power and waste management must be assessed within frameworks that extend beyond the technology itself. In the history of ideas, nihilism may appear to be a fundamentally human phenomenon. It may also be typical of modern society, as some philosophers point out. Thus, nihilism becomes the dominant characterisation of a technological era. As applied to the subject of this book, nuclear waste can be regarded as a necessary evil. On the one hand, nuclear energy was produced for reasons that appeared to be good at the time, and now the consequences must be dealt with. On the other hand, nuclear waste is a powerful symbol of a technological and nihilistic society that ploughs ahead without considering the future. The role of technology in modern society as a tool of breathtaking progress, as well as a source of enormous problems, deserves consideration.

Nihil is the Latin word for nothing. St. Augustine used it to describe non-believers. Philosophers began using the concept in the early nineteenth century and accused Fichte's idealism of being nihilistic. Nihilists were those who believed that nothing actually exists and who denied the reality of the values set forth by morality and religion. Starting with Russian literature, use of the concept accelerated in the latter part of the nineteenth century. Turgenev's *Fathers and Sons* (1861) describes nihilists as those who refuse to accept either authority or fixed values. Dostoevsky's frenetic tales brought criticism of nihilism to a wide circle of readers in both Russia and Western Europe. Critics in the 1880s and onwards linked the concept to Russian revolutionary movements, particularly those that sought to bring down tsarist rule by means of assassination. However, the Bolsheviks were also described as nihilists in the early 1900s.

Nihilism is associated with criticism of contemporary society in the 1890s and succeeding decades. The dissolution of traditional society – where people lived in rural communities, villages or small towns – was attributed to industrialisation, urbanisation and secularisation, which herded them into the restless metropolis. Factory work was mechanical and soulless. Time-honoured values withered away in the anonymity of the big city. Apostasy spread along with hopelessness. Christian faith lost its power of attraction, while secularisation robbed people of a cohesive worldview. The lack of principles and norms was seen as characteristic of the times.

This type of social criticism in German philosophy and thought led directly to the post-war analysis of technology and associated demands for a new concept of responsibility to address its consequences. The most important thinkers in this connection started writing during the interwar period. They had watched the initial war fever in 1917 dissipate under the devastation of endless trench warfare. The previous generation had experienced the faith in progress that swept over Germany after unification in 1871, the strong economic growth that spawned huge fortunes, the technological breakthroughs that spurred growing consumption. Turn of the century cultural life was subsequently characterised in contradictory ways. It was referred to as *la belle époque*, but also as decadent, pessimistic and effete (*fin de siècle*). Nietzsche criticized the optimistic spirit of the German Reich. He was read and referred to in the interwar period when his critique of contemporary society as nihilistic was well known. Of particular interest was his argument in favour of a nihilistic attitude towards old authority figures.

A critique of civilisation nourished by francophobia was cultivated in the Weimar Republic. The word civilisation was linked to Enlightenment ideas, positivism, mechanical thinking and atomistic individualism. Thomas Mann's conservative manifesto *Observations of a Non-Political Man* (1918) countered such trends with praise of German culture, which allegedly stressed cultivation and morality and served as an organic community that permitted individual development under a collective umbrella. Mann typically regarded culture and civilisation as two different types of spirits. Civilisation is 'spirit in the sense of reason, morality, doubt, enlightenment and finally dissolution, while culture is the life-sustaining and life-clarifying principle that organises and refashions the world on the basis of artistic vision'.[6]

Not all criticism of civilisation was conservative or opposed to democratic governance. After the war, Mann defended the new republic and parliamentary democracy.

The critics of civilisation represented divergent schools of thought. Max Weber devoted much of his writing to a description of the rationalisation process that Western society was undergoing with respect to both markets and public administration. His characterisation of the expansion of bureaucracy, as well as his gloomy conclusion that it placed people in an iron cage, was frequently cited. The Frankfurt School warned of perfunctory Enlightenment thinking that turned ideas and language into commodities. Claiming that the Enlightenment promise of liberty had been betrayed, representatives of the school criticised positivism as a scientific method and countered it with critical theory. Positivism was linked to capitalism and authoritarian tendencies in the political and popular culture of modern society. Criticism of civilisation by Theodor Adorno and Max Horkheimer lived in symbiosis with a pessimistic attitude that led them to draw drastic conclusions in light of Nazism and the Second World War: reason was liquidating itself.[7] Herbert Marcuse placed more emphasis on possible alternatives to contemporary tendencies. As opposed to Weber, he argued for critical rationality: modern technology could benefit freedom as well as support authoritarian patterns. Society could be structured in a wholly rational manner that offered individuals the opportunity to realise their potential for free action.[8]

The critique of civilisation often revolved around technology. Richard Coudenhove-Calergi, founder of the Pan-European movement, claims that 'civilisation has transformed Europe into a prison and most Europeans into slave labourers'. While he concedes that technology is required to liberate people from such slavery, it must be placed in the service of ethics. Mechanisation and standardisation destroy people's taste, creativity and natural instincts, poison their souls and turn them into machines.[9]

A cogent set of concepts arises when nihilism is used in criticism of civilisation. Mann has no doubt that his stance is anti-nihilistic. He praises Nietzsche's anti-nihilism in defending the German cultural tradition against the French orientation towards civilisation.[10] Hermann Rauschning's anti-nihilistic critique of Nazism in the 1930s and 1940s attracted a good deal of attention because he had been a member of the party for three years before moving to Switzerland in 1934 and ultimately to the United States. He regarded the Third Reich as a nihilistic revolution that had destroyed traditional German values. The Nazi doctrine of violence was an affirmation of nihilism, while the lack of ethical considerations in its politics was a degenerate form of philosophical nihilism. A total lack of norms took root under Nazi rule.[11]

An important step is taken when critics of civilisation identify nihilism as the force behind technological advances, degeneration and contemporary trends rather than machinery or the belief in progress. Nihilism is seen as a European phenomenon that developed along with nineteenth-century positivism, which stressed observable phenomena and paved the way for a denial of moral values and subjective experience. Because there was no longer a God to provide guidance, everything was permitted.[12]

Karl Jaspers and Hans Jonas play a special role in the discussion that follows. Both of them were phenomenological philosophers. Jaspers was a professor at the University of Heidelberg during the interwar period and at the University of Basel after the war. Jonas studied theology and philosophy at the University of Freiburg in the 1920s. Both men were close to Heidegger – Jaspers as a friend and colleague and Jonas as a student. They rejected Heidegger's involvement with Nazism and concluded that it was related to the decisionist aspects of his existentialism.[13] Jaspers and Jonas developed critiques of civilisation that use the concept of nihilism as a backdrop against which to describe the problems of contemporary society.

Jaspers uses the concept of nihilism in both his philosophical works and critical tracts. It appears in *Psychology of Worldviews* (1919), his first major work, which presents what he considers to be the fundamentals of human life. The book sets forth the idea of boundary situations with which he came to be most associated. According to this idea, both individuals and entire societies are constantly moving between places in which they feel at home and a state of dissolution. Dissolution and boundary crossing arise from the knowledge that there are other ways to live. Individual and social life are both part of an ongoing process that transforms the face of existence. Jaspers describes the emergence of worldviews that strive to avoid the uneasiness, suffering and pain that such change generates. He also looks at the growth of a pressing need for nihilistic dissolution. Nihilism in this sense is a fundamental force (as well as an idea or a principle) of spiritual life.[14]

In describing contemporary society, Jaspers often portrays historical changes in general terms – for instance, people have been uprooted and have begun to invest more hope in earthly existence than in the dream of an afterlife. The consequences are crucial to the contemporary worldview: people feel imprisoned in a transitory universe, which nourishes a sense of powerlessness. They are beset by the awareness that everything perishes, by constant questioning, by an endless whirlwind of self-deception.[15] In the spirit of Weber, Jaspers also takes a more sociological approach. Whereas Weber spoke of the disenchantment (in German, *entzauberung*) of the world, Jaspers uses the term de-deification (*entgötterung*) to describe the legacy of Protestantism and scientific thought. The result has been the rationalisation and mechanisation of production and organisation, as well as the triumph of methodical thinking. Jaspers emphasises that technology is extending its tentacles to the entire world. It squeezes its way into and schematises everyday life, making human interaction impersonal.[16] He characterises the modern world as the epoch of the machine that reduces individuals to cogs. Nations and cities, factories and shops, are bureaucratic machines that cannot see beyond the present. People lose their sense of past and future. The only thing that matters is their ability to operate the machine in the moment.[17]

In subsequent years, Jaspers writes that the Second World War has carried nihilism to its logical conclusion. He provides a basic definition that coincides with the one used in the nineteenth century. Nihilists have only temporary dogmas. They question and relativise everything. To them, there is no truth and all is permitted. Nihilists are driven by vitality and the lust for power.[18] The result was Nazism and the disaster of war.

> Today there are many forms of actual nihilism. There are people who appear to have abandoned every pretence of self, which seems to have lost

> all value. They stagger randomly from moment to moment, die indifferently and kill apathetically. But they live in the intoxicating idea of a quantitative existence, exchangeable with blind fanaticism, driven by fundamental, absurd, overpowering and evanescent emotions, and ultimately by the quest for instant gratification.
>
> If we listen to the words that are spoken, they serve as a concealed preparation for the science of death. Mass education makes us blind and unthinking, fanatically striving to master everything, including death and killing, ultimately taking even the genocidal war machine for granted.[19]

Thus, Jaspers identifies the fundamental features of nihilism and relates them to fanaticism and disrespect for human life, criticising a society in which the machine has taken over and people have resigned themselves to serving it. Nor can modern science escape his criticism. On the contrary, Jaspers regards science and faith in progress as the main threats to the post-war world.[20]

The threat of nuclear technology

Karl Jaspers's view of nuclear technology as a threat to the world is elaborated on in *The Atom Bomb and the Future of Man* (1958), in which he focuses on the consequences of nuclear weapons. He does not confront the consequences of nuclear waste. Citing scientific and engineering experts, he argues that the waste can be rendered harmless and placed in shafts deep below the surface of the earth.[21] Thus, he reveals a good deal of naiveté. But Jaspers sees the atom bomb in light of the technological age and the criticism of civilisation that he has previously advanced. What has changed is that the consequences have become so dramatic. New war-making capabilities, of which the bomb is the quintessential expression, have lent violence an entirely different dimension. That which was once a battle of conflicting interests now threatens to annihilate the human race.[22] People were previously able to kill themselves and others, even commit genocide. But now they have created a genuine threat to all life on earth. As a result, consideration of political conflicts must be transcended in favour of analysing the forces that bind together all social institutions. Jaspers argues that our ways of thinking must adapt to the new reality. The limited mindset of a philosopher or public official falls short when existence itself is at stake. Every individual is affected by the nuclear threat. A new ethos that proceeds from transpolitical motives is needed.[23]

Jaspers alternates between objective facts and philosophical reasoning. His solution is based partially on the Baruch Plan, which was proposed by the US government in 1946 and eventually led to the Nuclear Nonproliferation Treaty and IAEA. In accordance with the plan, all stages of atomic energy production (both uranium deposits and reactors) would be placed under the control of an international agency, which would have the right to monitor and inspect plants and production processes anywhere in the world. Jaspers posits that rational thought will avert the threat. In the spirit of Kant, he distinguishes between two ways of thinking. Human understanding is responsible for the mechanical approach of the technological age that has given birth to the atom bomb. Reason, which goes beyond simple understanding, is a fundamental, independent way of thinking capable of making judicious decisions. Reason holds the seeds of salvation.[24]

Jaspers proposes what he refers to as the principles of peace, which primarily involve the renunciation of violence as a means of resolving conflicts, to deal with the dangers posed by the atom bomb. He envisages a world in which laws and agreements are honoured, all countries waive their sovereignty and veto rights in international bodies, voting and majority decisions are complied with, news channels and public debate are honest and truthful, and human rights are broadly respected. In other words, he places his faith in 'soft' democratic values.[25] But Jaspers says that these principles are not enough. The more fundamental challenge is for human beings to 'change ourselves, our characters, our moral-political wills'. The ethics he demands is based on transpolitical motives and a generally moral attitude. Everybody must work towards such a change by creating peace in their own lives and passing it on to others.[26] A political ethics of peaceful intercourse is needed.[27] Such an ethics requires the courage to accept human limitations and to let go of our faith in technological progress.[28]

Responsibility as an alternative

The concept of responsibility is embedded in modern European history. Discussions of representative government advanced it as a political idea in the latter part of the eighteenth century. It was a philosophical idea in the nineteenth and twentieth centuries. Now it is integral to existentialism, phenomenology and neo-Kantianism. Largely as a response to the consequences of modern technology, the post-war era has given rise to a new concept of responsibility, particularly among German philosophers. This and the following section aim to identify a concept of responsibility which is relevant to incorporate into the seemingly intractable problem of nuclear waste management.

Assuming that critics of civilisation were correct in characterising twentieth century society as suffering under the yoke of nihilism, the next question is what to do about it. One obvious conclusion is that the constant inroads of nihilism dismantle the values on which power and authority are based, thereby creating new opportunities. All loyalties can be reconsidered. Heidegger pointed unabashedly to Nietzsche's discussion of the 'will to power'. Decisionist ideas, according to which those who act are able to seize power and initiative, are at the forefront. As Faust says in Goethe's rendition, 'In the beginning was the deed'.

Jaspers uses philosophy to describe the way out of the reductionism introduced by technology. He speaks of people's ability to dwell within themselves and develop their own personal assets. He writes of their fundamental being, which is the state of living with boundary situations. According to Jaspers, the meaning of life is to see such situations as an appeal to freedom and transcendence. Existential philosophy offers the possibility of finding an alternative to the dominance of technology.[29] Its promise of freedom does not provide people with a sense of calm. No heroes, prophets or demagogues carry the banner, and there are no battle lines. The transformation takes place in everyday life.[30]

Two words that Jaspers repeatedly uses in describing alternatives to unchecked nihilism are responsibility and future. He presents responsibility as the proper posture when faced with boundary situations, a third way beyond nihilism and the fear of change. The future represents the opportunity to choose freedom. The two words are embryos that presage his critique of nuclear energy.[31]

Before looking at the meaning of responsibility in criticism of nihilism and the implications for nuclear waste management, let us examine the history of the concept.

The word responsibility goes further back than either nihilism or legitimacy.[32] It arose during the transition from medieval to modern Europe, a period that historical research has charted quite thoroughly. The word was not part of the medieval vernacular. According to Germanic dictionaries, it was first used in the fifteenth century. Equivalent words appeared in English in the mid-seventeenth century, French in the eighteenth century and Italian in the nineteenth century.

The first known use of responsibility (*verantworten*) in German was synonymous with reply (*beantworten*). Grimm's dictionary gives the example of a prince who replies to a letter from the emperor. The word soon took on the meaning of pleading one's cause. Luther said that he penned his letters and other writings to take responsibility for his actions and beliefs. Responsibility in that sense might involve justifying oneself in a number of different venues – before sovereigns, theologians and congregants in Luther's case. The word came to connote defending oneself in a court of law and eventually emerged as a legal concept. Grimm's dictionary refers to a letter of 1665 taken from the archives of the principality of Weimar. The dictionary of the Swedish Academy gives many examples of responsibility as a legal concept in the eighteenth century. But the notion of moral responsibility to God also gained currency during that period.[33]

While the first transition in the history of the word responsibility is from reply to defending oneself before a legal or moral judge, the second transition is from credibility to a person or institution that can be held accountable. The *Oxford English Dictionary* ascribes the meaning of credibility to the word starting in 1642: 'The same Indentures were drawn also by the Kings Councell, in whose judgement and responsibility, the Vintners had reason to confide'. A prerequisite for accountability is that responsibility for a person, object or task be regarded as an obligation. As of 1737, the dictionary gives a number of examples in which the word accountability is used. *Verantwortlichkeit*, the German word for accountability, also appeared during that period. Thus, the word responsibility acquired a formal meaning that proceeded from the notion of holding someone accountable. In a similar vein, a business owner is always required to keep a book of accounts.

The formal meaning of responsibility reappears as a late-eighteenth-century political concept to describe various kinds of representative government. A Swedish councillor was responsible to the king and a British minister was responsible to the sovereign. Moral responsibility grew to be a secular, political concept that could be linked to a conviction of one type or another. Starting in the early nineteenth century, responsibility could connote an emotion. Grimm's dictionary gives examples of Romantic authors who used the word to describe a condition, a feeling from which it was impossible to escape. Thus, it could be associated with ideas of virtue and an inner sense of obligation.

Responsibility had the greatest impact on the nineteenth century as a political concept. It shows up repeatedly in the political writings of John Stuart Mill. He frequently looks at the division of responsibility when discussing representative government. It is obvious to him that liberty implies individual responsibility and that a developing society needs institutions to promote it. *On Liberty* argues that municipal autonomy requires civil servants to assume responsibility for ensuring that local affairs are conducted in the best possible way within the constraints of the law.[34] Around the turn of the twentieth century, Weber distinguishes between the responsibility of civil servants and that of a leading politician. Civil servants are not responsible in the sense that they carry out orders issued at a higher level, whereas the emperor is not responsible to anyone. The politician, on the other hand, cannot abjure responsibility for his actions. Responsibility defines his position.[35] Weber's distinction reveals the modern nature of his responsibility concept, linked as it is to the political system that took shape in the nineteenth century. His concept reflects a society based on legal and political individualism and is thereby closely related to citizenship and legitimacy.

Responsibility was first used as a philosophical concept in the nineteenth and twentieth centuries. Eighteenth-century philosophy discussed morality and ethics, as well as the status of free will versus law and determinism, in relation to responsibility. But the word itself was not used.[36] *The Metaphysics of Morals* (1797), Kant's last great work, speaks of obligation and imputation. Although that is rightly regarded as an ethic of responsibility, he does not use the word except in a footnote that appears in an appendix. Even then, responsibility has the older meaning of defending oneself.[37] While Mill freely uses responsibility in political and economic contexts, he is much more discriminating in his philosophical works. Only with existentialism does it become a fundamental philosophical concept. Kierkegaard's exploration of the ethical and religious life, as well as Heidegger and Sartre's analyses of being, all associate the concept of responsibility with the individual.

A new responsibility

Responsibility becomes a central concept for new ethical directions in the latter part of the twentieth century. It is consistently regarded in relational or socio-psychological, rather than individual, terms, as arising out of encounters with the Other. The focus is on defining responsible action and elaborating its distinctions. Furthermore, the horizons of responsibility are broader. *Responsibilities to Future Generations: Environmental Ethics* (1980) contains a number of important philosophical treatises from the English literature.[38] Meanwhile, German philosophers Hans Jonas and Karl-Otto Apel – otherwise very different thinkers – both argue that ethics must expand to embrace the entire planet, and they prescribe a collective sense of responsibility that includes the future as well. Their philosophies mutually reinforce the argument that nuclear waste management must consider these aspects of responsibility. But they also offer alternatives that point to different dimensions regarding the ethics of nuclear waste. This section will examine their concepts, and the next chapter will apply them.

Jonas is the critic of nihilism who takes the concept of responsibility furthest. Like Jaspers, he contrasts responsibility with nihilism. While Jaspers asks the big questions, Jonas devotes himself to the answers. He proceeds from a fairly general definition: 'The disruption between man and total reality is at the bottom of nihilism'.[39] He regards nihilism as an old phenomenon. In the early 1950s, Jonas completed the work on ancient Gnosticism that he had begun back in the late 1920s. He was originally struck by the similarities between Gnosticism and modern thinking. He subsequently concluded that Gnosticism was an old form of nihilism, which enabled him to more clearly understand the modern version as channelled by existentialism, primarily Heidegger. Like existentialism, the Gnostic doctrine of a God divorced from the world leaves human beings without a moral compass. But Gnosticism is not simply nihilistic; it also finds a purpose in eternal life. Modern nihilism is radical in that it offers no guidelines or objectives for human action: 'That only man cares, in his finitude facing nothing but death, alone with his contingency and the objective meaninglessness of his projecting meanings, is a truly unprecedented situation'.[40]

Jonas maintains that philosophy must find alternatives to nihilism, but he does not believe that existentialism fits the bill. *The Imperative of Responsibility* (1979), his magnum opus, argues that earlier philosophers had treated the concept in an overly restrictive manner. He obtains some guidance and assistance from religion, which extends the concept of responsibility beyond human life to include nature as well. The new dimensions of responsibility that Jonas presents include the need to go beyond the anthropocentric framework of previous philosophy. But he says that contemporary religion lacks the ability to fend off the nihilism – which is both powerful and vacuous, mixing extensive knowledge with ignorance about the direction of human existence – of technological civilisation. Only a new ethics can remedy the situation.[41]

Jonas presents a doctrine of human action in the face of modern technology. The enormous forces spawned by technological civilisation require a new ethics that must observe the cumulative impact in order to permit responsible, forward-looking action. A new ethics must incorporate the realisation that human life is rooted in global conditions and that any action can threaten its very existence. A suggestive but somewhat opaque sentence points to the essence of the dilemma: 'The gap between the strength of foresight and the power of action creates a new ethical problem'.[42] One interpretation is that human beings have great power to act that is unaccompanied by sufficient knowledge of the consequences. Thus, action must be based on an ethics that is equal to the challenges and threats of modern technology.

Given that scientific and technological progress was now threatening the future of the human race, he argues that a new imperative of responsibility is needed. Imperatives must be formulated that affirm the right of future generations to the planet. Like Jaspers, Jonas believes that the contemporary world demands an entirely new ethic. Traditional ethics did not look more than one generation ahead and focused on circumscribed societies – the city-state of antiquity or the modern nation-state.[43] But now humanity is faced with 'a growing sphere of collective action … the enormity of whose challenges requires an unprecedented dimension of responsibility'[44] for the entire biosphere, 'the global conditions of human life and its distant future, existence itself'.[45] The question of time horizons represents the most radical ethical transformation to which Jonas calls attention. The Baconian rationality that has led to unparalleled technological progress is incapable of assuming responsibility for generations to come. On the contrary, the fruits of technology are endangering the very existence of humanity. Jonas says that technology has unintentionally taken over, is galloping ahead at a pace that is harder and harder to control, and holds out a mirage of never-ending progress that can only result in universal disaster.[46]

Jonas advances an ethical imperative that will guarantee the integrity of future generations or avert threats to human existence. The implications of the imperative beyond actual survival are unclear, given that he also talks about 'genuine' human life and expressly rejects manipulation of people as biological beings. The imperative could also mean that future generations should be given the opportunity to lead decent social and cultural lives. Nevertheless, his goal is to establish a norm for public policy, not for individuals.[47]

According to Jonas, responsibility differs from ordinary rights and obligations. No reciprocity is demanded. The concept does not hold people accountable to others and is thereby neither legal nor political. His primary description refers to the spontaneous sense of responsibility that people feel towards children. Thus, its archetype is bestowed by nature.[48] The idea touches on the core of Jonas's thinking, which sees biological life as the basis for all philosophy and ethics. He wants to make a detour around the dualism between body and soul, instinct and will, that is so common in Western thought by finding a purpose in nature. Central to his philosophy is that life affirms and propagates itself.[49] However, he argues that there is a special, incontrovertible obligation to preserve the human race because of its ability to transcend nature. With their rational thought, capacity for reflection and free will, human beings are the highest expression of nature's purposiveness. The obligation does not lead to anthropocentric conclusions, given that the fate of humanity and nature are clearly intertwined. He says that human beings and nature have a symbiotic relationship and a common destiny.[50]

Assuming that there is a natural basis for such responsibility, an obvious objection is that human beings have created the very technological civilisation that threatens their future and that any remedy is doomed to fail. People may have an intrinsic self-destructive power that bodes ill for their survival. Jonas rejects such arguments. Action is certainly at the core of his thinking – nihilism is a human construct. But the problems that people have created carry the obligation to address them. Nihilism serves as the basis for establishing a standard of responsibility that is rooted in human existence itself, as well as the knowledge that humanity has acquired for destructive purposes.[51]

In Jonas's view, this sense of responsibility provides the basis for the optimum ethical posture, even in comparison with previous theories of morality. It is more profound than the concept of the greatest good that Western philosophy has handed down ever since Plato's eros and Aristotle's eudemonism, more important than Kant's respect for the law based on pure reason, and more sustainable than the existentialist subjectivism that ascribes the highest value to freedom of the self.[52] Above all, it is not simply formal, political or moral accountability. The latter stems from causal relationships embedded in human dealings and can lead to physical punishment, shame, the burden of sinfulness and the like. It is upheld by the principle of retribution. Responsibility, on the other hand, tries to predetermine human action with respect to specific values and objects. The ethics of responsibility for the future protects the integrity of humanity and of nature.[53]

While Jonas sees the responsibility that is vital to an ethics of the future as bequeathed by nature, he argues that there is also a kind of responsibility that derives from human freedom. The first kind of responsibility is inescapable, while the other is associated with free will. Responsibility bequeathed by nature is global, the foundation of social life. He clearly regards it as a transhistorical reality, a reflection of humanity's biological origins and sense of entitlement. It is in our bones like a primal phenomenon. On the other hand, he characterises the responsibility that is linked to free will as abstract and subject to choice. It requires specific interpersonal agreements.[54]

Based on his distinction between natural and contractual responsibility, Jonas offers two images to clarify his thinking. The first image is of a child, particularly an infant, who appeals to, and makes demands of, its parents and the adult world from a position of utter helplessness. A sense of responsibility for a being incapable of advocating for its own rights compels adults to protect the child. The responsibility extends to the social realm as the child grows older and requires education and acculturation. For Jonas, a child is the archetype and foundation of responsibility, as well as a universal source of experience. The second image is that of the statesman, for whom responsibility is a public matter. Responsibility does not begin at a specified point, but imposes itself on statesmen in particular situations. It originates from free will and is linked to society's institutional structure. Both parents and statesmen have a protective obligation that shapes their lives. They must maintain a sense of solidarity and identification with those to whom they are responsible. Continuity and far-sightedness are vital in both cases. Society is to be preserved, while children are to learn speech and other skills. Both are to be incorporated into a historical community. Tomorrow is a constant presence for both parents and statesmen.[55]

This is the decisive component of responsibility in Jonas's view. Human existence is created and shaped in time. Responsibility is a moral complement to our ontological position. In that respect, he does not explicitly refer to responsibility as bequeathed by nature, but emphasises the time aspect instead.[56] While proceeding from the truism that human beings live in a historical context – that parents feel a sense of responsibility for their children, and statesmen for citizens – is persuasive, it can also be called into question.

One possible objection to Jonas's argument is historical. Not all cultures have assumed that children are entitled to life and nurture. The point at which a child is regarded as a human being worthy of preservation changes from one period to another. Thus, what Jonas refers to as responsibility bequeathed by nature is not inalienable. One explanation might be that free will temporarily intervenes and damages it. Such perversions would not upend the idea of parenthood as the origin of all responsibility. Children could still serve as archetypes.

A more serious objection can be levelled against Jonas. He regards nature not only as a foundation, but as an inexorable, seething ferment that brews a sense of responsibility. From such a point of view, it is difficult to make out where historical, changing responsibility takes over. For instance, the very use of the word responsibility is linked to modern legal, political and economic individualism. Key aspects of the responsibility that Jonas speaks of are also associated with a globalised world and the time horizons that have emerged from the intersection of technology, its consequences and opportunities for collective action. By proceeding from transhistorical conditions, he makes it difficult to discern changeable and culturally distinct phenomena.[57]

An alternative would be to look at the question through the other end of the binoculars and regard the new type of responsibility as a goal rather than a starting point. What would be decisive in that case is the ability to address contemporary threats while including both the global community and future generations. Karl-Otto Apel, who writes in the post-Kantian tradition and has collaborated on discursive ethics with Habermas, offers another way of understanding responsibility in relation to such issues.

Apel does not speak of nihilism in contemporary society, although he refers in the 1980s to the 'global crisis of techno-scientific civilisation'.[58] Like Jonas, he identifies the need for an ethics of responsibility as a response to modern science and technology. The environmental crisis and nuclear armament are the clearest manifestations of the predicament. Natural resources are scarce and nuclear weapons can destroy the world.[59] Thus, he speaks of the 'responsibility of our times' and the need for a 'macroethics' or 'global ethics' that can guide humanity.[60] Like Jonas, he argues that a new historical situation has arisen and forced the human race to assume collective moral responsibility.

Although Apel praises Jonas for bringing the need of a new universal ethics to the fore, he is deeply critical of the way that Jonas's principle of responsibility focuses on preserving humanity and the conditions for its survival without leaving room for a concept of progress and improved living standards.[61] Apel carries his objection to its logical conclusion by linking Jonas's principle to the Social Darwinist view that humanity can more readily survive if some Third World populations starve to death. He makes it clear that Jonas does not think that way, but asserts that his principle of responsibility fails to erect obstacles to such a solution.[62]

Apel's critique concerns the natural basis of Jonas's philosophy.[63] Whereas Jonas grounds his ethics on biology and nature, Apel proceeds from reason: people see the necessity of a new type of responsibility in their capacity as rational creatures. Thus, he places himself in the Kantian tradition. As rational creatures, people can also demand equity on the same terms wherever they live, and even for future generations. For Apel, this concept of equity points forward and towards social progress. He extends the principle of responsibility to include preservation of human life *and* dignity.[64]

Apel sees what he calls communication communities as a means of implementing such a principle of responsibility. Discourse and human beings as discursive creatures are the pillars of his responsibility. He repeatedly emphasises that the way in which discussions are conducted reveals a kind of ethics. People enter into a discussion under particular historical circumstances on the basis of specific human inclinations and interests. Meanwhile, people stake a claim to an ideal community by participating in a discussion. They assume the existence of a communication community based on the norm that everyone is accepted as an equal partner who shares responsibility for addressing the problem. One basic ethical norm is that consensus can be reached by means of argumentation. That is the prerequisite for entering into a discussion in the first place. Thus, there is a meta-norm that transcends situational norms and resides in human reason.[65] Apel does not try to defend his thesis by looking at the past, but takes his examples from modern society and clearly reflects the basic norms that are associated with a democratic, constitutional state. While citing Kant's assertion that the norms underlying policy-making, legislation and administration must be subject to public discussion in order to achieve legitimacy, he does not examine the history of the idea.[66]

The difference between Jonas and Apel is illustrated by their approaches to the responsibility of elected officials. Jonas proceeds from ancient Greek philosophy and the discussions by Solon, Lycurgus, Pericles and other lawgivers,[67] whereas Apel considers the role of modern office holders. He rejects Weber's notion that elected officials are responsible to their constituents only and that ethics should be relegated to the private sphere. He seeks to erase that distinction by basing ethics and responsibility on the elements of reason that are inherent to communication. Apel asserts that elected officials have an ethical responsibility. The tension between a specific political system, with all the conflicts and private interests that entails, and an ideal communication community is particularly challenging. He says that responsible officials should promote the long-term ascendancy of 'the basic norm of conflict resolution through argumentative consensus building'.[68]

Apel bases his concept of responsibility on an ethical rationality that he carefully separates from the institutional approach to creating legitimacy. The institutional approach involves strategic action that proceeds from calculated self-interest, as manifested in economism and politics. Ethical rationality stems ideally from discussions that are made possible by shared rules and norms.[69] The sense of responsibility ultimately comes from an awareness of the gap between the current and the ideal communication community, as well as the insight that improvement is necessary and possible.[70]

The strength of Apel's argument is that his concept of responsibility includes equity between present generations. Thus, he provides an important alternative to Jonas. Furthermore, humanity is not regarded as an eternally abstract category. Responsibility implies an ideal of equity and the possibility of moral progress. All this goes beyond Jonas's philosophy.

Notes

1 Fjodor Dostojevskij *The Brothers Karamazov*, Constance Garnett's translation, p. 1726.
2 IAEA, http://www.iaea.org/cgi-bin/db.page.pl/pris.charts.htm.
3 Benjamin K. Sovacool 'Critically weighing the costs and benefits of a nuclear renaissance', *Journal of Integrative Environmental Sciences*, 7:2, 2010, pp. 105–23.
4 Sovacool op. cit. Andrew Blowers 'Why Fukushima is a moral issue? The need for an ethic for the future in the debate about the future of nuclear energy', *Journal of Integrative Environmental Sciences* 8, 2011, pp. 73–80.
5 Per Högselius 'Spent nuclear fuel policies in historical perspective: an international comparison', *Energy Policy* 37, 2009.
6 Thomas Mann *Betrachtungen eines Unpolitischen*. Frankfurt am Main: S. Fischer Verlag, 2002, 1983 [1918], p. 185: 'Geist im Sinne der Vernunft, der Sittigung, des Zweifels. Der Aufklärung und Endlich der *Auflösung*, während Kultur im Gegenteile des künstlerisch organisierende und aufbauende, lebenerhaltende, lebenverklärande Prinzip bedeute.'
7 See Max Horkheimer 'Förnuftets slut' in John Burill (ed.) *Kritisk teori – en introduktion*, Göteborg: Daidalos, 1987, p. 352. First published as 'End of Reason' in *Studies in Philosophy and Social Science* 1941.
8 Herbert Marcuse 'Några samhälleliga konsekvenser av den moderna teknologin' in John Burill (ed.) *Kritisk teori – en introduktion*, Göteborg: Daidalos 1987, pp. 363, 142. First published as 'Some Social Implications of Modern Technology' in *Studies in Philosophy and Social Science* 1941.
9 Richard N. Coudenhove-Calergi *Revolution durch Technik*, Wien: Paneuropa Verlag, 1932, pp. 11ff, 71–75. Citation p. 11: 'Die Zivilisation hat Europa in ein Zuchthaus verwandelt und die Mehrsahl der Europäer in Zwangsarbeiter'.
10 Mann op. cit. p. 102.

11 Hermann Rauschning *Die Revolution des Nihilismus: Kulisse und Wirklichkeit im Dritten Reich*, Zürich: Europa Verlag, 1938, pp. 49, 146ff.
12 See Martin Heidegger *Nietzsche: Europäischer Nihilismus*, Gesamtausgabe Bd. 48, Frankfurt am Main: Klostermann, 1986 (1961), pp. 2–16.
13 Richard Wolin *Heidegger's Children: Hannah Arendt, Karl Löwith, Hans Jonas, and Herbert Marcuse*, Princeton: Princeton University Press, 2001.
14 Karl Jaspers *Psychologie der Weltanschauung*, Berlin 1922 (1919), pp. 219, 280–84.
15 Karl Jaspers *Die geistige Situation der Zeit*, Berlin: Sammlung Göschen, 1931, pp. 5ff, 14.
16 Jaspers op. cit. pp. 16ff, 26f.
17 Jaspers op. cit. pp. 29ff.
18 Karl Jaspers *Der philosophische Glaube*, München: R. Piper 1948, p. 103.
19 Jaspers op. cit. pp. 119f: 'Heute gibt es mannigfache Gestalten des faktischen Nihilismus. Es sind Menschen erschienen, die scheinbar jedes Selbstsein preisgegeben haben, denen nichts Wert zu haben scheint, die im Zufall von Augenblick zu Augenblick taumeln, die gleichgültig sterben und gleichgültig töten, – die aber zu leben scheinen in den berauschenden Vorstellungen eines quantitativen, in blinden Fanatismus auswechselbarer Art, getrieben von elementaren, sinnfremden, übermächtigen und doch schnell verbrausenden Affekten, und schließlich vom triebhaften Genusswillen des Augenblicks.'

'Hören wir den Worten zu, die in diesem Treiben gesagt werden, so wirken sie wie eine verschleierte Vorbereitung des Sterbenkönnens. Massenerziehungen machten blind und gedankenlos, um im Rausch der Hingabe zu allem fähig zu werden und am Ende den Tod und das Töten, das Massensterben im Maschinenkampf als selbstverständlich hinzunehmen.'
20 Jaspers op. cit. pp. 128ff.
21 Karl Jaspers *Die Atombombe und die Zukunft des Menschen*, München: R. Piper Verlag, 1958, p. 20.
22 Jaspers op. cit. pp. 82f.
23 Jaspers op. cit. pp. 21–31, 48.
24 Jaspers op. cit. pp. 33, 39–48. Citation p. 53: '... der grossen Vernunft, die mehr ist als blosse Verstand'.
25 Jaspers op. cit. pp. 40–46.
26 Jaspers op. cit. p. 50f. Quotation on p. 50. '... unsere sittlich-politischen Willen müssen wir verwandeln'.
27 Jaspers op. cit. p. 485.
28 Jaspers op. cit. pp. 53, 281f.
29 Jaspers *Die geistige Situation der Zeit*, pp. 133f.
30 Jaspers op. cit. pp. 150, 157f.
31 See Jaspers *Psychologie der Weltanschauung*.
32 The remainder of this section is based primarily on *Deutsche Wörterbuch von Jacob Grimm und Wilhelm Grimm*, Leipzig, 1854–1971, *Historisches Wörterbuch der Philosophie*, Basel: Schwabe, 1971–2007, the *Internet Encyclopedia of Philosophy*, the *Oxford English Dictionary*, Oxford: Clarendon 1986 and *Svenska akademins ordbok*, Stockholm 1903.
33 See Google Books.
34 John Stuart Mill *Considerations on Representative Government*, London 1867, *On Liberty*, London 1859.
35 Max Weber *Wirtschaft und Gesellschaft: Grundriss der verstehende Soziologie*, Tübingen: Mohr Siebeck, 1980 (1922), p. 833.
36 Richard McKeon 'The development and the significance of the concept of responsibility', in Zahava M. McKeon (ed.) *Freedom and history and other essays: an introduction to the thought of Richard McKeon*, Chicago: University of Chicago Press 1990.

37 Immanuel Kant *Die Metaphysik der Sitten,* Frankfurt am Main: Suhrkamp Verlag, 1956 and 1977 (1797), p. 627.
38 Ernest Partridge *Responsibilities to Future Generations: Environmental Ethics*, Buffalo, NY: Prometheus Books, 1981 (1980).
39 Hans Jonas 'Gnosticism, Existentialism and Nihilism', in *The Phenomenon of Life: Towards a Philosophical Biology*, Evanston: Northwestern University Press, 2001 (1966), p. 234
40 Jonas op. cit. p. 233.
41 Hans Jonas *Das Prinzip Verantwortung: Versuch einer Ethik für die technologische Zivilisation*, Frankfurt am Main: Suhrkamp 1984 (1979), pp. 26ff, 57f, 99f.
42 Jonas op. cit. pp. 22–30, citation p. 28. 'Die Kluft zwischen kraft des Vorherwissens und Macht des Tuns erzeugt ein neues ethisches Problem.'
43 Jonas op. cit. p. 28ff.
44 Jonas op. cit. p. 31.
45 Jonas op. cit. pp. 33f.
46 Jonas op. cit. pp. 201f.
47 Jonas op. cit. pp. 36f.
48 Jonas op. cit. pp. 85f.
49 Jonas op. cit. Chapter 3, particularly pp. 128f.
50 Jonas op. cit. pp. 157, 245ff.
51 Jonas op. cit. 230–36.
52 Jonas op. cit. 165–71.
53 Jonas op. cit. 171–76.
54 Jonas op. cit. 178–83.
55 Jonas op. cit. 182, 196ff, 234–41.
56 Jonas op. cit. p. 198.
57 Peter Kemp *Das Unersetzlische – eine Technologieethik*, Berlin: Wichern Verlag, 1992.
58 Karl-Otto Apel *Diskurs und Verantwortung: Das Problem des Übergangs zur post-konventionellen Moral*, Frankfurt am Main: Suhrkamp 1990 (1988), p. 177.
59 Apel op. cit. pp. 17, 23, 180f and 247ff.
60 Apel op. cit. pp. 42, 176.
61 Apel op. cit. pp. 42f, 183ff.
62 Apel op. cit. p. 196.
63 Apel op. cit. p. 45.
64 Apel op. cit. pp. 184f.
65 Apel op. cit. 46ff, 67, 202.
66 Apel op. cit. 206.
67 Jonas op. cit. pp. 42f.
68 Apel op. cit. pp. 256–61, 260: 'die ideale Grundnorm der Konfliktlösung durch argumentative konsensusbildung entbunden'.
69 Apel op. cit. pp. 55–63.
70 Apel op. cit. pp. 141 et allem.

5 The uncomfortable responsibility

This chapter analyses and discusses responsibility in relation to nuclear waste, now applying the conceptual frame from Jonas and Apel. It evolves the argument that nuclear waste management should consider ethical responsibility including equity and rights of future generations. The chapter includes an exhortation for nuclear waste management to maintain a historical perspective, and concludes with a discussion of whether deep geological disposal is a feasible solution.

From an ethical imperative to an ethical question, and back again

What is there about modern society that makes ethics such an important ingredient of legitimacy? The obvious answer is that the vast time horizons involved require new thinking about ethics. As Jonas demonstrates in a number of different ways, political responsibility entails an entirely new time horizon. The issue of spent nuclear waste can be recapitulated from that point of view. Important to note is that large quantities of waste have already been generated. Looking at the coming decades and the next few centuries (the medium term), the constant production of new waste is pretty much a given. Whatever happens, there will be radioactive waste on or under the earth's surface for thousands of years. Even the medium-term perspective requires the development of ethical norms that will be valid for a long time. The market norms that have evolved over the centuries illustrate such a process. Given that nuclear waste will remain in close proximity to communities for at least another century, it is starting off at a similar point. Because it will affect people and their social environment, it will be subject to policy decisions. The difference is that waste must be stored for thousands of years – thus, ethics has to serve in a monitoring and surveillance capacity. The following discussion will begin with the short-term and medium-term perspective. But the development of norms during that period must seek to formulate the imperatives that the long-term perspective requires. This is the topic of subsequent discussion.

A number of time horizons must be related to the question of responsibility. One horizon is based on Jonas's belief that humanity has the same type of responsibility for all future generations, thereby returning to the notion of an ethical imperative. This time horizon is nothing but eternity. Another time horizon is found in the current principle that each country and generation is responsible for managing its own nuclear waste. That involves a time horizon of approximately thirty years. As previously mentioned, this principle is of dubious value. In the first place, international ownership and technological progress make it difficult to argue for exclusive national responsibility. In the second place, the waste that is produced at any particular time will be around for many generations to come. Both horizons emerge as static when related to nuclear waste.

Jonas's ethical imperative and the ethical principle of nuclear waste management are vulnerable to Apel's objection that they ignore the possibility of progress. Apel concludes that Jonas cannot limit his principle of responsibility to the preservation of humanity but must append a concept of equity and dignity.[1]

Apel teaches that principles exist in a process of communication and appeals to the possibility of progress. Communication and progress will logically include negotiations. We will have nuclear waste for a long time to come and we can be sure that the development of norms for nuclear waste management will be subject to negotiations and agreements for many generations to come. One conclusion is that the ethical imperative represents an idealistic position. The decisive objection to the ethical imperative is that it is non-negotiable, that it is posited as a fixed, inflexible goal. Not even when nuclear power has been phased out and is no longer regarded as a useful source of energy, while all its waste is in repositories where radiation will eventually return to safe levels, is that kind of ethical rigidity appropriate.

Given that the fundamental ethical principle has lost its relevance, the question arises as to whether it is reasonable to have such a principle at all and, if so, how it should be formulated and what its underpinnings should be. What can replace the failed principle of the 1980s and 1990s?

The thesis of this book is that Jonas's ethical imperative (as well as Jaspers' principles of peace) must be supplemented. Apel furnishes the concepts of equity and dignity. Moreover, in constructing an ethic of nuclear waste, Jonas can also be supplemented with an ethic of the Other.

An exploration of various ethics of technology by Peter Kemp, a Danish philosopher, clarifies the issue further. Kemp proceeds from Jonas's imperative of responsibility. His reading of Jonas highlights the importance of responsibility that embraces future generations. Stressing the threats that technology poses to humanity, he mentions nuclear weapons, Chernobyl and similar phenomena. The overall problem is 'that the science and technology that became popular because they would protect life against the threat of nature are now about to make life more risky than the state of nature itself'.[2] Of course, things aren't that simple. The repercussions of technology were perceived as threatening even during the Industrial Revolution. But now the magnitude of the danger seems much greater.

Objecting that the imperative of responsibility treats the human race as an abstraction, Kemp modifies Jonas's viewpoint by emphasising the role of solidarity with real people. Jonas's impersonal ethic must be rooted in an approach, found primarily in the writings of Emmanuel Levinas, that considers the suffering, difficulties, experiences and wisdom of individual human beings. According to Kemp, the lessons gleaned from this ethic of the Other are a prerequisite for an impersonal ethic of the type that Jonas advocates.[3]

Kemp's argument sheds light on deep geological disposal of spent nuclear fuel and high-level waste for hundreds of centuries. Such a solution may be regarded as a radicalisation of Jonas's imperative of responsibility for future generations. The extraordinarily long time horizon inherent to the solution means that all generations to come – both *Homo sapiens* and its successors – must be considered.

Given that people will be living in proximity to nuclear waste for at least several more generations, the equity ethic of Apel and Kemp's/Levinas's ethic of the Other appear to be reasonable. Such an ethic would have to uphold the rights of the next few generations while promoting equity, democracy and other values in the face of the challenge that internationalisation poses to the concept of national responsibility. An ethic of the Other requires discussions, negotiations and choices about the construction of an equitable society and democratic order. An impersonal ethic ignores the concepts of equity and democracy.

Kemp is careful to point out that Levinas's ethic of the Other must be limited to prevent it from encompassing other demands and to maintain targets over the course of many generations. In his view, Jonas's imperative of responsibility serves exactly that purpose. One conclusion is that nuclear waste management requires both an impersonal ethic and an ethic of the Other.

Speaking of ethical principles in the context of nuclear waste management appears to be a problem. Kemp's principle of vulnerability is not analogous to Jonas's imperative of responsibility. In the case of nuclear waste, abandoning the effort to formulate an ethical imperative and speaking of an ethical question instead would be a highly reasonable approach. Such an orientation would depart from a simple attempt to establish fixed targets in favour of queries about how responsibility is to be taken, who is to take it and the way in which norms are to be developed.

Nevertheless, refraining from the consideration of any ethical principles for the management of material whose radiation is hazardous in both the medium and long term appears to be an unreasonable demand. Thus, the formulation of ethical questions must be linked to the ethical principles associated with long-term horizons. Such a realisation refocuses on the need of better articulating an ethical imperative.

Imperatives

Kant's description of the way that reason uses imperatives to create order among various sources of knowledge is a good place to start at a time when the issue of nuclear waste management is disintegrating into different national and political discourses. According to Kant, an ethical imperative may be understood epistemologically, such as by establishing a heuristic function for knowledge acquisition.[4] Ordering and systemising a major issue like the consequences of nuclear energy and waste must proceed from a regulative principle. An ultimate imperative is required to link short- and long-term horizons while addressing a host of problems and solutions.

But can the concepts of imperative as advanced by Kant and Jonas be reconciled? While Kant's imperatives are ideas independent of and beyond empirical reality, Jonas regards them as experientially based. Kantian imperatives are accessible through reason, whereas Jonas's imperatives proceed from shared experience. For both philosophers, an ethical imperative is a universalist argument. Kant says that all people possess reason. Jonas, on the other hand, is an Aristotelian who maintains that children's instinctive appeal to the adult world constitutes the experience on which the imperative of responsibility is based.

The difference may not be so dramatic if agreement is reached on what the imperative should be. But the existence of a universalist reason independent of history is difficult to claim. Constructing an imperative related to nuclear waste management in the light of historical conditions would appear to be more reasonable.

The most important lesson to be learned from the history of the concept of imperative is that it serves not only as a premise but as a bearer of power that provides it with a normative function. Integral to Kant's thought is that an imperative also permits the establishment of a normative premise. In line with his philosophy, it may be argued that ethical imperatives are required for shaping various approaches to nuclear waste management, providing principles for lawmakers to follow and framing laws. Proceeding from Kantian thought in this way while citing Jonas, who maintains that all previous philosophers wrote of their own times, is paradoxical. Arguing that they were interested in the 'immediate surroundings of action', Jonas urges a different perspective, which surveys the future that technological civilisation holds out. Kemp makes a crucial contribution to the discourse by maintaining that the ethic of the Other and Jonas's imperative of responsibility must be regarded as complementary. Important to point out, however, is that Kant is also instrumental in describing an ethic that assumes responsibility for human beings of the future. The paradox may be resolved by distinguishing between Kant's practical philosophy and the fundamental role that he assigns to imperatives. Even as the views of Jonas and Kant are reconciled, various time horizons must be isolated and a general imperative formulated for the long term. But there are ethical issues that must be dealt with in the short term as well.

'When principles are invoked, common sense flies out of the window', Doris Lessing writes in her autobiography. Nuclear waste management must be treated as an ethical issue in the short term, replete with constant conflicts and demands for renegotiation. Principles should serve as a target and point to the long-term perspective. But principles cannot easily resolve immediate conflicts of interest, as they are empty of substance from a policy perspective. As such, a principle is either indecipherable or subject to arbitrary interpretation.[5] In the short term, an ethic merges with issues that require policy decisions – it can set targets but not prescribe concrete measures. Based on ethical considerations of responsibility for future generations, it is fully possible to advocate deep geological disposal for thousands of years.[6] However, similar considerations may lead to the conclusion that storing waste above ground for the next few centuries would be a better option.[7] The problems associated with geological disposal will be discussed shortly.

Jonas stresses the importance of distinguishing between imperatives and utopia. He draws a line when it comes to the principle of hope as enunciated by Ernst Bloch. In three large volumes, Bloch assembles examples of hope as a fundamental principle of human life. Forward-looking hope explains much more than the retrospective approach taken by psychoanalysis and provides a needed complement to Marxist class analysis, with which he generally sympathises.[8]

According to Jonas's critique of Bloch, utopian thought tends to portray people as one-dimensional creatures who all fit into a single mould. Convinced that there is no fixed human nature, he promotes ambiguity and flees the assumption that people are inherently good or evil.[9] In his review, an ethical imperative that insists on responsibility for future generations cannot be regarded as utopian.

A long-term ethical imperative concerns responsibility for human life. But, as Kemp argues, such an imperative must be rooted in an ethic of the Other. Furthermore, it requires a global perspective far beyond the imperative of national responsibility that reigned earlier. In other words, it must be highly general, focused more on a target than on specific instructions for action. A troublesome dichotomy arises here: the need for both an impersonal imperative of responsibility that focuses on future generations and an ethic of the Other. While the impersonal imperative can obscure contemporary inequities, the ethic of the Other may lose sight of responsibility for generations to come. Once more, a conceivable resolution is a combination of a Kantian reason-based imperative and Jonas's empirically based imperative: the ethical imperative must be regarded as eternal while conceding that it can be framed only tentatively and can be subject to perpetual scepticism and renegotiation. It should be formulated as if it applied to the future even as it is rooted in the experience currently available. Thus, the imperatives of the impersonal ethic cannot be reduced to a fixed universal reason independent of contemporary interests and agreements.

Safety and equity

As suggested by Jonas and Jaspers, safety can be considered as a basis for ethical action when enumerating the dangers posed by technological civilisation and nuclear technology.

As suggested by the above discussion of Apel and Kemp, equity is also a factor to consider. Similarly, Habermas argues that genetic technology raises questions about what it means to be a human being and live the good life. Referring to John Rawls, he asserts that equity in modern society is closely associated with the right of individuals to carry out their particular life projects. Similarly, nuclear technology raises the classic philosophical question of equity. A society that wants to offer both equity and individual freedom has a profound need for an ever-expanding supply of energy. The resulting waste, which threatens life and the biosphere if not isolated, brings the issue of responsibility to the fore.

Responsibility must be rooted in the present and the future, the personal and the impersonal. It must address safety issues to preserve the conditions for biological integrity, as well as equity issues to protect human dignity. Thus, responsibility consists of four components.

Jaspers, Jonas and Apel demand safety and integrity for present generations.
Jonas and Apel appeal to safety and integrity for future generations.
Apel stresses the quest for equity among current generations. Kemp highlights the need for solidarity.
Apel's demand for equity also includes future generations.

Responsibility for nuclear waste is partly a matter of safety and partly a matter of equity. It should comprise both present and future generations. Thus, the responsibility that is being sought consists of these components.

Waste management must be safe for the human beings who are alive now. They should not be exposed to radiation due to negligent waste transport or management.
Waste management must be safe for future generations. Radioactive substances must be isolated from the biosphere until they have been rendered harmless. The difficulty is in deciding when radiation has returned to low levels. Questions arise concerning what should be regarded as normal or acceptable radiation. How these questions are answered will determine the time horizon to be specified when isolating the substances.
Waste management must be equitable for current generations. The weaker version of this component is that producers and consumers of nuclear energy are responsible for ensuring that the burden is not placed on those who derive no benefit from it. The strong version is that nuclear energy be regarded from the perspective of global equity such that its use in industrialised countries frees up other resources for emerging economies, or that emerging economies boost their energy supply by means of nuclear power while industrialised countries assume responsibility for waste management.
Intergenerational equity must be ensured. The simplest version of this component is that those who are alive now must not take advantage of nuclear energy to improve their living standards while placing the burden of

managing waste on those who live after them. That responsibility may include payment by current nuclear energy users into a fund set aside for defraying the costs of waste management. Both Sweden and the United States are already pursuing such an approach. A more daunting challenge associated with that responsibility is the extent to which nuclear energy will be used inequitably for the next seven or eight generations, leaving the waste for four thousand subsequent generations to manage.

Ocean and space dumping are not seriously discussed as options for dealing with nuclear waste. They are not safe for either present or future generations. Recycling technologies that generate plutonium lack feasibility for similar reasons. No transmutation technology is available. Such technology would radically reduce waste and shorten the required storage period to a couple of thousand years but would also spawn a market for the transport of spent nuclear fuel, thus increasing the hazards that present generations face. The option of deep geological disposal is being developed in Finland, Sweden and the United States, and is under serious consideration in Canada, Germany and Britain.[10]

Deep geological disposal, whether of spent nuclear fuel or high-level waste, appears to be a good way of fulfilling the responsibilities implied by the first and third components. The solution satisfies the demands of the first component by protecting contemporary society from waste and permitting continued nuclear energy production. How long radioactive nuclides can be isolated from the biosphere remains an open question. Kristin Shrader-Frechette has identified several methodological flaws in the geological and technological studies of Yucca Mountain. She argues that the geologically unstable mountain poses unacceptable disposal risks.[11]

Deep geological disposal addresses the demands of the third component in that nuclear energy producers like Sweden use coal and oil less, while one country's repositories can store the waste generated by another country. Meanwhile, deep geological disposal technology can be exported to other countries with suitable bedrock.

The second and fourth components raise problems that deep geological disposal cannot solve. The second component has two difficulties to surmount. First, nobody can know how nuclear waste capsules will be affected by seismic shifts through the millennia. Estimating the probability that the biosphere will be protected long enough is hostage to a number of assumptions. But even if physical protection works for the many thousands of years required, no technology can prevent human encroachment. If waste can be deposited deep below the surface of the earth, it can also be retrieved out of error, greed, curiosity or malice. Nobody knows whether generations of the distant future will possess the knowledge required to manage nuclear waste. To paraphrase Shrader-Frechette, deep geological disposal creates greater uncertainty for future generations than keeping waste above ground and under control.[12]

The fourth component is also problematical when it comes to deep geological disposal. If radiation starts to leak because something goes wrong with waste management or the capsule breaks, and knowledge is lacking about how to fix the problem, many people will fall ill. Given that their numbers will accumulate for many generations, no fund will be able to either cover the costs or compensate for the suffering.

Responsibility for nuclear waste is further complicated by the fact that it may be either unchanging or subject to recurring negotiation. Each of the four components can be regarded as both established principles and negotiable issues. Nuclear waste management must deal with this paradox. Extensive negotiations are required to determine how and where waste is to be stored, as well as who is going to bear the burdens. However, negotiations to determine a safe way of managing nuclear waste can also undermine public trust by implying that absolute standards are not being applied. The deep geological disposal solution has encountered such difficulties. As discussed in Chapter 2, popular protest stopped test drilling because it was perceived to be unsafe. The industry tried to circumvent the issue by arguing that bedrock could be selected even if it was not perfectly suitable. The implicit admission that repositories would not be optimally located set the stage for new, stronger objections. Given social realities, the deep geological disposal solution raises more questions than it answers.

The solution also poses new problems when the responsibility for future generations is considered to be negotiable, or even diminishing over time.

A schematic analysis of responsibility must also address the question of how waste management decisions are to be made. Such decisions are based on moral standards, legal regulations and political negotiations. Moral standards constantly influence regulations and policy while giving rise to ethical principles that stake a claim to general applicability. Policy decisions should ideally be made in connection with negotiations on responsibility that take ethical principles into consideration. The judiciary institutionalises responsibility for nuclear waste.

To be viewed as legitimate, nuclear waste management must be ethically sustainable, legally valid and politically acceptable. The above discussion makes it clear that the deep geological disposal solution cannot claim to be ethically sustainable. Nor are there any other solutions that can vie for a full-fledged legitimacy. Thus, Shrader-Frechette offers the most sensible solution: keep the waste within reach and monitor it until a better solution is available.

The need for temporal dimensions

The above emphasis on equity in relation to future generations raises the question of whether the equity dimension of responsibility is negotiable. A theory propounded by scholars of religion Mikael Stenmark and Carl Reinhold Bråkenhielm implies that such equity is indeed negotiable: our responsibility for each future generation is less than for the generation before it. The theory poses the crucial issues of intergenerational equity in a provocative manner.

With respect to nuclear waste, the theory claims that our responsibility for welfare and quality of life is strong for the next few generations but progressively weaker for more distant ones. One version of the theory sets the initial limit at 150 years. A weaker principle of equity states that current actions must not prevent generations that live during the subsequent 150 years from satisfying their needs. The theory posits a minimal principle of equity after that, requiring only that the possibility of life on earth not be entirely threatened. The upshot of the theory is that the current generation is free to use nuclear energy, but must assume responsibility for spent nuclear fuel and not place the burden on future generations. The strongest argument for a diminishing principle of equity is the difficulty of empathising with those who will live after one's great-great grandchildren. Only five or six generations can be alive at the same time.[13] The concept is clearly reconcilable with the notion that responsibility for the yet unborn consists of their safety and of not threatening life on earth. The concept of long-term responsibility is consistent with Jonas's thinking. The approach proceeds from a sense of general equity or responsibility that extends beyond the next few generations. The problem with the theory of diminishing equity is that it attempts to draw a line in the sand. It justifies the exploitation of scarce sources of energy as long as they are rationed such that a few more generations can benefit from them.[14]

This theory can be defined with the support of Apel's concept of responsibility. The thesis is just unsustainable if it claims that members of the current generation are responsible only to those with whom they have close affinity. In that case, the affluent would have no responsibility for people in other economic brackets. The European welfare state would forfeit its legitimacy. The theory of diminishing equity also appears abstract and idealistic, ignoring the sense of responsibility that future generations may have for their successors on the basis of their own assumptions. Determining how far into the future responsibility extends is an extraordinarily difficult task. In reality, people always feel responsible for those who will live a hundred years after them. The exact points at which responsibility weakens and ultimately becomes minimal are very hard to pin down. The future challenges that responsibility will face and the evolution of attitudes towards equity are impossible to foresee.

But even more objections can be raised. Many societies and communities have a sense of affinity for much longer than a couple of centuries. Nationalists often outdo each other in harking back to the distant roots of their particular nation. Christians cite emotional affinity with a group of people who lived two thousand years ago. Social narratives create a sense of community over long periods of time. Historical experience and discourses represent the primary objection to the thesis of diminishing equity. By the same token, people's understanding of their responsibility for nuclear waste is a function of how they view their place in history.

Kemp argues that narratives are needed if comprehensive ethical systems are to evolve.[15] When an ethic must take future generations into consideration – as is the case with nuclear waste management – the historical narrative assumes particular importance. The following is one example.

According to Thomas Mann, the young Goethe read and was fascinated by the story of Joseph in *Genesis* and suggested that it be turned into a major work. Taking up the gauntlet, Mann wrote a four-part novel that traces Joseph's fate as he is sold to Egypt-bound traders, becomes the Pharaoh's viceroy and ultimately saves Jacob's tribe from seven years of famine. From Mann's remarkable perspective, the parlance, religions and cultures of the contemporary world are conspicuously absent, while the human predicament is omnipresent. In no sense does Jacob live at the beginning of time; he follows many previous generations. Egypt is the land of wonders when Joseph the shepherd boy arrives. But its civilization, more than a thousand years old, finds itself at a crossroads between old and new ways of thinking, its priesthood having lost much of its power, its peasants and proprietors seeing their land surrendered to the Pharaoh's growing supremacy. Mann demonstrates the constant presence of human and historical origins, thereby providing a time horizon that extends across decades, centuries and millennia.

How people view their position in history is decisive to the issue of nuclear waste management. The idea is simple – human beings are as capable of empathising with their descendants as with their ancestors. The same kinds of narratives that link contemporary and ancient societies can be projected into the future. Direct encounters with future generations are impossible. However, narratives establish a mechanism for developing ethics of the Other (Levinas), vulnerability (Kemp) and equity (Apel) that include the unborn. Knowing that there is a past permits identification with the experience it has generated and people who have long since died. Employing imagination to create or assimilate historical narratives makes people more capable of assuming responsibility for those who come after them. By the same token, a view that human history is irrelevant leads to negligence of the rights and interests of future generations. In other words, legitimate nuclear waste management must retain a historical perspective. Universalism requires temporal dimensions.

The theory of diminishing equity can also be examined in the light of John Rawls's *A Theory of Justice* and the idea of an original position, according to which wealth and social resources are allocated on the basis of decisions by people who wear a veil of ignorance. The parties are equal, have a sense of justice and can decide what is best for them. But they know nothing about their own particular wealth, income or social status. Nobody can tell how any of them will benefit or be harmed by social or natural contingencies. Rawls assumes that the parties will allocate the assets among themselves such that each of them will tend to receive an equal share. Thus, natural reason represents a strong imperative for justice.[16]

Intergenerational resource distribution can be examined in the same way. If the veil of ignorance covered the century that the parties lived in, and the century that their children will be born in, what rights would the parents regard as inalienable? Would the rights differ if their children were to live in the next generation, 100 years later or 400 years later? The obvious answer is that the rights would be the same regardless. Thus, an intergenerational theory of equity can be based on universalist dimensions.[17]

The final question is what other areas are affected by the concept of diminishing justice. The idea is wholly inapplicable to social development. Both local and national politics are largely devoted to ensuring eternal progress in terms of transport, communication, health care, education and other services. Responsibility for future generations is a guiding principle. Conflicts may arise between various objectives – motorways can harm the environment, and so on. But the focus is always on the future. Although officials may be ignorant or incompetent with respect to the consequences of their decisions, they continue to look ahead. Not only the ideologies of liberalism, socialism and social democracy (which are based on faith in progress), but conservatism and environmentalism share this perspective.

The desire to create a better future is one of the most fundamental features of Western thought. From the beginning of modern times, the accumulation of capital has always been forward-looking. A number of theories have emerged since the early twentieth century to the effect that the ability to refrain from instant gratification is the basis of all future capital formation.[18]

The idea of diminishing responsibility appears to be a construct designed to lend theoretical legitimacy to plans for implementing a deep geological disposal solution. Borrowing from the words of Giorgio Agamben, diminishing responsibility might be described as a way of putting future generations under a state of emergency. As long as ideas, principles and grounds for legitimate nuclear waste management are related to the present generation, safety and equity are squarely addressed. But such considerations fly out the window when future generations are left to fend for themselves.[19]

Is deep geological disposal a feasible solution? A discussion

Objections to deep geological disposal have, perhaps surprisingly, been raised by Swedish and Finnish advocates of the KBS-3 method. On the one hand, they were firmly convinced that they had found a reasonable, technologically sound solution to the problem of nuclear waste. But no matter how completely the waste was banished from the face of the earth, they realised that encroachment could always imperil the solution.[20]

Opinions vary about whether deep geological disposal is an acceptable solution. However, many reputable scientists and engineers are favourably disposed to the model.

A basic objective of deep geological disposal is to be rid of the problems associated with nuclear waste – out of sight, out of mind. Based on the Yucca Mountain project, Kristin Shrader-Frechette describes the concept as an attempt to bury uncertainty. According to her, geologists do not know enough about how bedrock behaves over long periods of time. Relatively short-term measurements cannot predict what is going to happen for tens of thousands of years in the future. Making uncertainty invisible does not improve the situation. On the contrary, deep geological disposal poses the risk that future generations will dig up the waste out of ignorance or for malevolent purposes.[21] One response to this argument is that the bedrock at Yucca Mountain is less stable than the Baltic Shield or the Canadian Shield. But that leads to another crucial objection, this one non-geological in nature.

The author of this book remains unconvinced that deep geological disposal is a sustainable solution. Experts have raised technological objections. But for the sake of argument, let us assume that the method is defensible from an engineering and scientific point of view, that geological and technological knowledge can ensure safe storage until the waste is no longer hazardous. Then shouldn't construction of the repositories begin immediately? The answer is that, regardless of the scientific pros and cons, strong reasons for not doing so remain.

Dimensions of knowledge other than technological and scientific models and empirical tests – i.e. past experience and data – are relevant at this point. What becomes immediately obvious is that it is never possible to predict how future generations will act. Societies develop and change. The contemporary world places great faith in progress, but knowledge has lost ground during certain periods of history. The decisive objection to deep geological disposal must be raised on the basis of such awareness. What people who live centuries from now will do with the repositories built by the present generation cannot be known. The energy they contain, or the copper capsules themselves, might be regarded as precious bounty even though technologies to manage the risks are lacking. People may excavate in the bedrock without being aware of either the waste or the concept of nuclear energy. One proposal has been to include warnings about the hazards associated with the repositories. One challenge that presents itself right away is how to devise signs and symbols that civilisations of the distant future will understand. The most persuasive objection is that any such warning will entice fortune seekers. Whenever there has been a presumption that pyramids and other burial sites have contained valuable objects, the spontaneous impulse has been to open them up.

This means that the option of deep geological storage that is retrievable is not a solution for managing the nuclear waste, as it carries the same dangers of being forgotten as the alternative of a final repository. It is also a way of hiding the waste away, of following the 'out of sight, out of mind' precept.

But isn't nuclear waste more secure in deep repositories than intermediate storage? Perhaps, but it can be very well protected and still under human control without being buried. Swedish waste, for instance, is already well protected against encroachment, stored 45 metres down in bedrock where neither planes nor bombs can strike it. The advantage of leaving it there for the foreseeable future is that it remains under human supervision and expert monitoring. Such considerations might take a different form in other parts of the world. Not all countries are likely to build repositories of sufficient quality.

Thus, deep geological disposal will probably not reduce the risk of radionuclides leaking into the biosphere.

Furthermore, the relationship between nuclear waste and nuclear weapons bears examination. Because the countries that reprocess spent nuclear fuel obtain access to plutonium, deep geological disposal removes such fuel from plutonium production. Meanwhile, the nuclear powers will continue to produce plutonium regardless of whether deep repositories are built. Other developments are needed before nuclear disarmament can become a reality. The most important risk factor is that reprocessing and trading in reprocessed material multiplies opportunities for illegal production of nuclear weapons and the acquisition of nuclear capability by new countries. In other words, international monitoring of nuclear technology and spent nuclear fuel must be stepped up and internationalised far beyond the present level. Implementation and acceptance of deep geological disposal in a country like the United States might be regarded as an argument for reprocessing as a means to safely store the plutonium products that remain after recycling.

Intermediate storage is clearly not a final solution. The presence of waste at that level represents an imperative for society to seek better alternatives than simply hiding it and soon forgetting all about it. One problem is that accessible waste can be regarded as a recycling opportunity and a resource for plutonium production. The objection has been raised that the hazardous material would be more readily accessible. However, waste that is not recycled is difficult to handle – its radioactivity protects it against theft and unlawful use. The objection can be countered by the assertion that mechanisms exist for monitoring nuclear waste, and that preventing spent nuclear fuel from being reprocessed needs to be a top priority in any case. Ever since President Carter decided in 1977 not to take the reprocessing route, the view that recycling should be limited has gained currency and practice has followed suit. But voices have been raised in recent years that the United States should reprocess more in order to extract additional energy. Upholding the principle of limited reprocessing, thereby thwarting plutonium production and trading in spent nuclear fuel, is of vital importance. The benefits of extracting additional energy are not worth the risks posed by reprocessing.

But what happens if current social institutions break down? The answer is that such a scenario has already unfolded during the nuclear era. Even as the Soviet Union collapsed, control of nuclear weapons was maintained and new agreements were signed to deal with the situation. An important lesson can be learned from that period. Keeping hazardous material in view creates a reason for holding society together, either to sustain existing institutions or to create new structures.

So, we must preserve the social order in order to prevent the monster that we have conjured up, like the spirit in the bottle or the genie in the lamp, from destroying us completely. Nuclear waste management is necessary for safety reasons, but it can also be regarded as a series of rituals that contribute to community cohesion and, paradoxically, to peace and secure coexistence.

The argument of this book is to wait, not to run into the questionable solution of deep geological disposal, and to keep developing the various alternatives. But how does this jive with the 'polluter pays' principle? Who should pay for the waste management if not the ones producing it? This objection requires comment. Deep geological disposal fits very well with, and can even be regarded as a logical consequence of, the polluter pays principle. But the dignity of the principle is much less than the quest for responsibility. We must ask if the polluter pays principle should be allowed to infringe upon responsibility for the waste. If an ethically sustainable solution is unavailable, this simply has to be acknowledged. At the end of the day, management of nuclear waste can only be required of a company, but the responsibility remains on the legislators.

Nuclear waste should remain in intermediate storage where it is safe and can be monitored, until better solutions appear or the twilight of the nuclear era arrives. Current knowledge will survive during that period and alternative waste management methods will have time to emerge. Meanwhile, resources should be devoted to further development of methods for management, including transmutation and various techniques of disposal.

If no other solution arises, the risks of intermediate storage can still be weighed against those of deep geological disposal. But there is no reason to make that decision now; as long as nuclear technology is used to extract energy, intermediate storage is the wiser approach. Thus, my conclusion is that the effort to identify the best solution should continue.

Notes

1 Karl-Otto Apel *Diskurs und Verantwortung: Das Problem des Übergangs zur postkonventionellen Moral*, Frankfurt am Main: Suhrkamp 1990 (1988), pp. 184f.
2 Peter Kemp *Das Unersetzlische – eine Technologieethik*, Berlin: Wichern Verlag, 1992.
3 Kemp op. cit. pp. 106–11.
4 Immanuel Kant *Kritik der reinen Vernunft*, Frankfurt am Main: Suhrkamp Verlag, 1956 and 1974 (1781–87), p. 320.
5 Stanley Fish *The Trouble with Principle*, London: Harvard University Press, 2001 (1999), p. 5.
6 Mikael Stenmark and Carl Reinhold Bråkenhielm 'Nuclear Waste, Ethics and Responsibility for Future Generations' in *Nuclear Waste: state-of-the-art reports 2004*, Stockholm, SOU 2004:67.
7 K. S. Shrader-Frechette *Burying Uncertainty: Risk and the Case against Geological Disposal of Nuclear Waste*, London: University of California Press, 1993.
8 Ernst Bloch *Prinzip Hoffnung*, Frankfurt am Main: Suhrkamp, 1985 (1959).
9 Hans Jonas *Das Prinzip Verantwortung: Versuch einer Ethik für die technologische Zivilisation*, Frankfurt am Main: Suhrkamp 1984 (1979), pp. 333ff.
10 Urban Strandberg and Mats Andrén (eds) *Journal of Risk Research* 7–8/2009.
11 Shrader-Frechette op. cit. Barry Solomon 'High-level radioactive waste management in the USA', in *Journal of Risk Research* 7–8/2009, pp. 1009–24.
12 Shrader-Frechette op. cit.
13 Stenmark and Bråkenhielm op. cit.

14 Stenmark and Bråkenhielm may erect other obstacles to the use of scarce sources of energy, but that goes beyond this discussion.
15 Kemp op. cit. pp. 85ff.
16 John Rawls *A Theory of Justice*, Oxford: Oxford University Press, 1971, § 4.
17 Rawls op. cit. § 44.
18 Mats Andrén *När den nya nationalekonomin kom till Sverige: marginalismen, den österrikiska skolan och Knut Wicksell*, Göteborg: Göteborgs universitet, 1994.
19 Giorgio Agamben *Undantagstillståndet*, Site: Lund 2005 (2003), p. 79.
20 Gunnar Gustafson 'De tekniska principerna bakom det svenska slutförvaret för använt kärnbränsle – KBS 3' in Mats Andrén and Urban Strandberg (eds) *Kärnavfallets politiska utmaningar*, Hedemora: Gidlunds förlag, 2005.
21 Shrader-Frechette op. cit.

6 Moral culture and the formulation of norms

Conditions for the formulation of norms

Norms for managing nuclear waste cannot be formulated on the basis of the current idealistic view of technology and the problems it creates. Such idealism combines the unabashed notion that technology can solve any problem with a vaguer but widespread brand of optimism. It is closely tied to the idea of never-ending economic growth and wealth formation. The primary objection to this idealism is that the ability of technology to solve every conceivable problem is highly uncertain and that any attempt to do so is likely to create others.

Nor can norms be formulated based on the kind of elementary realism that demands a reduction of greenhouse gases in response to climate change, additional energy resources or the use of nuclear power to protect national or European security. Such realism advocates a biosphere policy, a joint global effort to meet these challenges. An oversimplified approach is usually taken that highlights various facts, each of which contains a large measure of truth. But this elementary realism lacks an overall perspective that considers distribution of wealth, conflicts of interest, equity, ethics and democracy.

The formulation of norms that can address the uncertainties associated with nuclear waste management of the future must describe the issue in a way that is relevant to political, legal and ethical discourses alike. It must proceed from problems that society actually faces, as well as the challenges posed by demands for sustainable legitimacy. Social life is bound up with materialistic questions and considerations, conflicts of interest, political issues and economic factors. It raises questions about the proper distribution of wealth and approaches to creating a better world. However, the formulation of norms must look beyond the existing social order as well. It is part of a discourse that examines ethical and democratic issues in relation to sustainable legitimacy.

Below is an overview of an approach to relating nuclear waste management to an ethical discourse. The aim of this chapter is to emphasise the relationship between a moral culture and the nuclear waste issue. The description must be tentative. The prerequisites for constructing norms are reflected in two steps. First are general conditions with respect to perceptions of society and humanity ('moral culture', 'understanding contemporary society', 'cosmopolitanisation is a prerequisite for the formulation of norms'). Second are two sections on the idea of progress, which is of specific relevance to the nuclear waste issue ('various images of progress', 'deep geological disposal and the idea of progress').

Moral culture

Central to the Kantian view of humanity is the fundamental obligation to seek perfection. Rather than a homage to individualism, the focus is always on collaboration within a social structure. People are obligated to cultivate their natural abilities. The farmer learns to plough and the professor to research or teach. Furthermore, people are to act on the basis of the ethical maxims prescribed by the categorical imperative. The only incentives for such acts that Kant specifies are the sense of obligation and the satisfaction derived from obeying it. He stresses the existence of an internal moral culture.[1]

Kant associates moral culture with universal reason, which the bourgeoisie possesses. A moral culture with such a foundation is difficult to conceive of in modern society. However, the phrase carries the connotations on which the formulation of norms for nuclear waste management must be based.

Modern philosophy emphasises the difference between establishing and applying norms.[2] The difference is exceptionally challenging when it comes to nuclear waste. In the first place, the ethical principle that has been in place since the early 1980s dictates that the burden of nuclear waste not be passed on to future generations. But the time horizons involved are enormous and no permanent solution has yet been found. In the second place, the nuclear-energy-producing countries have different geographic prospects for managing their waste. However, that gap can be bridged.

To start with, nuclear waste management must be coupled to an ongoing discussion about moral responsibility for the negative consequences of technology. The responsibility is accountable not only to laws and public policy decisions, but to the issues of moral philosophy that arise in connection with long-lived radioactive waste that will affect future generations.

Given Kant's idea of universal reason, the translation of moral culture to modern philosophy inevitably leads to the view that there are certain eternal values. This does not necessarily imply the rejection of particularistic or relativistic norms, but at least an insistence that general principles exist. In any case, nuclear waste must be managed in keeping with some kind of ethical universalism.

The development of ethical universalism in the Aristotelian tradition examines the aspects of life that recur in all cultures and identifies the attitudes they necessitate. Amartya Sen lists a number of functions that people must satisfy in order to live a meaningful life. As biological creatures, they need proper nutrition and health. As social creatures, they thrive on self-respect and integration in the community. The list goes on and on. Martha Nussbaum, a moral philosopher, enumerates eight common human traits that all cultures share: mortality, the body, pleasure and pain, cognitive capability, practical reason, early infant development, affiliation and humour. Everyone gradually realises that they are destined to die. All people have the same basic corporeal needs. In every culture, the body is the source of ideas about pleasure and pain. People have an inherent ability to obtain and develop knowledge, as well as to plan and make decisions about their lives. They share the experience of having been at the total mercy of others as infants, and they have bonded to one another in friendship or love. Finally, all cultures make room for humour, albeit in very different ways. In the spirit of Sen and Nussbaum, philosopher Tore Nordenstam suggests that there are 'universal ethical parameters'. But he says that the problems themselves – as opposed to norms, ethical rules or proposed solutions – are the parameters.[3]

In other words, ethical questions, rather than ethical principles, are universal. However, nuclear technology and waste must be governed globally on the basis of global norms. In this case, common ethical parameters must be translated into universal norms. That requires going beyond the neo-Aristotelianism that Nordenstam invokes back to Kant's question of moral culture and universal reason.

Proceeding from Kant, norms can be said to attain full legitimacy once deemed to be universally sustainable. He addresses the issue of how to construct yardsticks of ethical assessment based on individual life situations, as well as shifting social opportunities and challenges. Apel, who places himself in the Kantian tradition, argues the existence of a problem that Kant fails to resolve from the contemporary point of view: how to harmonise individual and collective claims to resolving the ethical issue of the good life.[4] He proposes a principle of universalisation, which is not a universally valid ethical precept but an ongoing discourse that continually clarifies the restrictions to which people and institutions should be subject in order to ensure realisation of the good life. Most important is that the discourse be ongoing, pragmatic and sensitive to the problems and issues that society faces.[5]

The formulation of universal norms for managing nuclear technology and waste should originate from this kind of practical discourse. The norms must start from both Aristotelian and Kantian principles. The basic foundation for the formulation of norms arises from contemporary society, its discourses and institutions, as well as historical conditions. Any norms that are established must be accompanied by awareness of the need for sustainability over the many epochs that nuclear waste requires, while acknowledging that they will always be subject to renegotiation and reconstruction. Given the demands that responsibility makes, norms include the safety and equity of future generations. In other words, the formulation of norms is forward-looking but the process of constructing them is related to the challenges that are known at the time. Thus, development of an ethics that stretches into the future must proceed from an ethics of the Other and current moral parameters. Even when starting with history and the contemporary world, however, ethical principles must be the aim.

In contemporary society

The nuclear waste management issue begins at the intersection of the private and public spheres. Questions are raised about control and ownership rights, as well as the monitoring role of the authorities. The next juncture is the broader issue of national versus international responsibility. International collaboration is needed between producers, public agencies and non-governmental organisations. Globalisation in the areas of economics, politics, communication, media, culture and civil society is central to the process. Today's challenges to the sovereignty of the nation-state come into play. All of the issues that arise from locating nuclear waste management at these two intersections are closely related to various interests. Ownership rights are in flux, although there is a clear tendency towards privatisation. Nevertheless, the actions and decisions of national governments will largely shape the future management of nuclear waste.

The third intersection is between technology and democracy. The question, the broadest problem of all, is what the driving force for social growth and development should be. The issues that come up involve knowledge, democratic control, norms, decision-making, advocacy, equity and solidarity in a globalised world. Nuclear waste management must address the issue of technology and legitimacy. That is the only conceivable basis for sustainable legitimacy.

Nuclear technology places special demands on society. Both nuclear weapons and nuclear power for peaceful purposes require a large measure of security and monitoring at the international level. The IAEA reflects the fact that a new context has emerged to handle implementation of nuclear technology. Thus acknowledgement of the new context is essential.

What is the larger context of the nuclear waste issue? Some characteristic features of contemporary society are relevant here. One such feature involves new types of complexity. Social scientists have tried to conceptualise it ever since the early 1980s, and in different ways expressed their sense that society was undergoing a fundamental change. Beck argued that modern technological developments entailed risks that were global in nature and subsequently elaborated a concept of cosmopolitisation.[6] Later on, Habermas described and analysed elements of what he referred to as the post-national constellation.[7] John Urry amplified the concept of complexity as a means of understanding contemporary society.[8]

This is not to imply that society had previously been transparent and easy to understand. Every era has its unique qualities and nebulous tendencies. Many historians and social scientists have been driven by the challenge of grasping the particular society they lived in. Hegel's *The Phenomenology of Spirit* (1806), Marx's *Capital* (1867) and Weber's *The Protestant Ethic and the Spirit of Capitalism* (1904) were all motivated by this desire. Habermas, Beck and Urry are among the many social scientists in recent decades who have developed concepts for the analysis, interpretation and comprehension of contemporary society.

How can contemporary society be described? Representative democracy has lost some of its ability to create legitimacy for public policies and commitments, one of which is nuclear waste management. The triumph of neo-liberalism has placed a greater emphasis on private ownership, not least in the nuclear energy sector. The fall of Communism and the integration of Europe have moved boundaries. Globalisation has transformed economics, politics, communications, culture and many other spheres. Nuclear waste management as a national policy-making issue framed by national structures has been under increasing stress since the 1970s.[9] Increasingly multi-level societies have made more room for local interests and governance,[10] which is now an important factor in several nuclear waste programmes. As a result, the nation-state must find new ways of solving problems. Among them are privatisation and the emergence of multinational corporations, which defy national sovereignty when it comes to energy and many other areas.[11] What will happen to national responsibility for nuclear waste when the nuclear energy sector is dominated by international corporations is an open question. Can we expect the nuclear industry to keep the regulatory boundary lines between nuclear fuel and nuclear waste separate? Would the nuclear industry not prefer to include the possibility for trading nuclear waste in the global and entirely commercial resource cycle for fissionable products?[12]

In the past few decades, Habermas and David Held have both emphasised the need to shape global public opinion, the formulation of transnational norms that stem from cosmopolitan values. That is particularly true of nuclear technology, which requires global norms if it is to be globally regulated. The challenges posed by nuclear weapons, the peaceful use of nuclear energy and the management of nuclear waste all demand such an approach. Because nuclear waste management is caught between the standardisation of transnational technology and the framework and conditions of nation-states and the constraints of public opinion in democratic society, it is saddled with major difficulties.

A 1984 speech by Habermas to the Spanish Parliament refers to 'a new impossibility of taking an overall view'. He links that situation to the fact that the welfare state found itself in a blind alley and utopian thought had lost steam. Contemporary society was no longer oriented towards the future (a typical ingredient of modern thought).[13] The new impossibility of taking an overall view that Habermas speaks of may be regarded as part of the background to formulating norms for managing nuclear waste. Faced with such impossibility, three approaches are possible. The first approach is longing for a society that is clearly structured, able to shape norms and make decisions by means of elementary, easily understood processes. Whether any society has ever resembled that kind of paradigm is another question altogether. Discussion of national responsibility and the ethical principle may be manifestations of such longing, reflecting a desire for a sovereign nation-state that can deal with the issue. In other words, the focus is on the norms that are formulated in the nation-state and that impact its decision-making institutions.

The second possible approach to addressing the impossibility of taking an overall view is to seize control of it. Given the global society we live in, that kind of action would require strength and undisputed authority. Proposals for global regulation of the nuclear fuel cycle have been advanced on a number of occasions. The Baruch Plan of 1946 led to the establishment of the IAEA. Mohammed El-Baradei, former Director General of the IAEA, suggested the establishment of a global nuclear fuel bank in 2005. The Soviet Bloc adopted a common nuclear fuel cycle. The Euratom Treaty represents the EU's attempt to ensure a supply of nuclear fuel. This approach does not consider the need for global norms, which are crucial in building legitimacy for the management of nuclear waste.

The third approach to dealing with the impossibility of an overall view is to argue that it promotes citizen participation and democratic processes. A multi-level society cannot make decisions overnight but is dependent on complex negotiations that are transparent and embrace different interests and objectives. In the case of nuclear waste management, the levels are local, national, regional and global.

Cosmopolitanisation is a prerequisite for the formulation of norms

In the short and medium term, ethical issues associated with nuclear waste management will revolve around equity between different parts of the world, the rich and the poor, as well as responsibility to future generations, the promise and future shape of democracy, and the role of technology in social and personal life. As Helmut Fleischer began to articulate two decades ago, such ethical dimensions must be examined without fixed imperatives as part of a wide-ranging discussion that is conducted from the vantage point of global citizenship.[14]

Unambiguously linking nuclear waste management to the imperative of assuming responsibility implies the necessity of acting as global citizens. Three reasons can be cited.

The first reason is the danger represented by nuclear weapons. The technology for producing fuel to be used at nuclear plants and for reprocessing the spent fuel can also be used to develop nuclear weapons. Reprocessing is critical because it is the source of plutonium production. If the responsibility for reprocessing remains with sovereign nation-states, each country is free to develop its own nuclear weapons capability and thereby pose a threat to the entire world. The second reason is that nuclear waste can fall into the wrong hands and be used for terrorist or other criminal purposes. The third reason is that nuclear plants are part of the international energy network, which raises questions about global equity and the ability of developing countries to industrialise.

The problems are largely addressed at the national level. Individual countries decide to develop nuclear energy, either with the help of industrial players or on their own. They assume responsibility for managing nuclear waste.

They formulate policies to secure their future energy supply. In democratic countries, such choices are made with the participation and consent of the governed. But reactors and nuclear energy production raise issues that extend to the global arena. In each case, global citizens must act.

If norms for nuclear energy production and waste management are to be effective and broadly applicable, they must be formulated at a transnational level. Measures are needed that promote the eradication of poverty and support the kind of global equity that marginalises neither current populations nor future generations. The role of public policy and the citizenry must be protected and expanded. Norms are needed that determine what is to be publicly and privately administered, while informing the kind of social planning that can regulate the accumulation of global capital As Habermas constantly emphasises,[15] global opinion must be shaped on the basis of transnational norms that proceed from cosmopolitan values – particularly when it comes to technologies that develop in transnational constellations and that irreversibly change human life and social organisation over the long term.

The concept of cosmopolitisation introduced by Beck is key in this connection. While other thinkers tend to speak of cosmopolitan values as desirable or necessary for global regulation, Beck says that the world is locked into the process of cosmopolitisation whether it likes it or not. People cannot escape it but must learn to live with it. Cosmopolitisation, which pervades contemporary reality, is not the same thing as economic globalisation. The reality that cosmopolitisation describes is both political and international in scope, involving bottom-up changes to lifestyle, employment, travel and other patterns of behaviour. 'The historically irreversible fact [is] that people from Moscow to Paris, from Rio to Tokyo, have long since been living in really existing relations of interdependence.' Beck sees dangers in this situation, as well as the opportunity for global affinity and cosmopolitan responsibility.[16]

Philosopher Peter Sloterdijk says that the world is experiencing a moral climate change, that distances and borders are growing less important in the wake of faster transport and communication: 'No longer do people have to live together in order to be connected. They need not be related to feel something for each other. They require no common illusion to show solidarity with each other. They do not have to personally witness something before helping each other. The sum of these relationships is what I call long-distance neighbourhoods.'[17] The quotation is enlightening, even though Sloterdijk advocates the expansion of nuclear energy for the purpose of reducing carbon dioxide emissions, thereby representing a nihilistic approach. Thus, a wide range of social observers see the world as increasingly cosmopolitan.

The nuclear waste issue is a quintessential example of the global interdependence of contemporary societies, which reveals the more profound and ancient truth that human beings rely on each other for their happiness and survival. However, norms are always formulated in tandem with assumptions about the cultural characteristics of humanity. Whether tacit or clearly articulated, such assumptions determine how a society is understood and explained at a particular time and place. Social and ideological conflicts stem partly from the collision of differing perceptions about the human condition. If global norms are to be established, these underlying perceptions must be brought to the surface and integrated with each other.

Can 'cosmopolitanisation' and the 'cosmopolitan world' be reconciled with contemporary individualism? Researchers are fairly unanimous in their belief that contemporary society is characterised by a type of individualism that is either new or stronger than ever before. They have varying assessments of this; some talk about opportunity and freedom, others about a new kind of uncertainty and inability to address common problems. The most prominent form of individualism is the description and ideal of human action as self-interested and profit-maximising. That such a view of the human condition has become prescriptive reveals the conformist tendencies of contemporary society.

There are other individualisms that serve as alternatives to the concept of economic man. According to Habermas, the individual portrayed by neo-liberal ideology is remarkably diminished. What is lacking is the notion of a moral person whose decisions are based on an awareness of the common good, as well as the idea that citizens of a republic participate in the legislative process on an equal footing.[18]

Both the political and moral views of the human condition must play a central role in any sustainable system for developing norms that can be applied to nuclear waste management. The concept of cosmopolitan citizens suggests a global presence in the present that proceeds from democracy and equity. The idea of moral individuals suggests a sense of responsibility for both current and future generations. Thus, norms are required that insist on the obligation of acting as moral individuals and cosmopolitan citizens. Benjamin's angel has to turn around to see the people of the future and meet their gazes. A sense of solidarity must emerge that extends to those who are not yet born.

Various images of progress

Jonas says that humanity faces an uncertain future while holding onto a deeply rooted faith in progress.[19] The dichotomy reflects a fundamental conflict in contemporary society and the issue of nuclear waste management. Progress is central in contemporary formulations of norms regarding society and the future. Thus, the implications of the concept of progress need to be explored. The history of ideas has exhaustively analysed the concept. Johan van der Pot has meticulously examined various views of technological progress in *Die Bewertung des technischen Fortschritts: Eine systematische Übersicht der Theorien* (1985), from which three types of arguments can be identified that may be applied to the nuclear technology and the nuclear waste issue: optimistic, pessimistic and the redefinition of progress.[20]

First, the optimistic type of argument concerning progress denies nuclear threats, or accepts them but embellishes them with faith that technological progress will ultimately rescue humanity. A specific application of nuclear technology stirred the imagination of humanity during the 1950s. Nuclear power seemed to furnish an inexhaustible source of energy. Fantasies of space travel abounded. More earthbound visions focused on the prospect of environmentally friendly energy sources. As opposed to coal and other fossil fuels, atomic energy was considered to be pure. Fossil fuels were black, while atomic power was labelled 'the white energy'. The dream of pure energy was nourished by the concept of progress, which was associated with the notion of endless economic growth. Growth is seen as integral to both personal prosperity and the national interest. Advocacy of nuclear technology for civilian purposes has often relied on this argument, as argued by Andrew Blowers only months after Fukushima:

> The pressure to support new nuclear gives emphasis to contemporary needs, to the production of energy, investment in jobs, development of big technology to increase supply. The risks from emissions, accidents, from nuclear waste, even from proliferation become acceptable in an ethical discourse which speaks the language of progress, economic growth and modernisation. It is in these circumstances that an affirmative answer is given to the ethical question of whether radioactive releases can be justified.[21]

The problem posed by nuclear waste was previously regarded as temporary, a state of affairs that would only last until technology could solve it. However, technological progress both promises a better world and threatens to destroy it. Second, in the pessimistic view, the nuclear threats may destroy humanity and perhaps all life on earth. This is the argument of Jaspers. The pessimistic argument treats humanity as a collective category without considering the differences that arise from inequity and other factors. As a result, the quest for an ethical principle such as a Jaspers' peaceful orientation assumes central importance. But while Jaspers only speaks of the atom bomb and the inconceivability of its use, the problem for the present generation is also the very real existence of nuclear generation and nuclear waste. The question of universal values must be posed in a new way in order to confront the peaceful use of nuclear power and the waste it generates. Given the consequences of nuclear energy, contemporary observers related it to the need for universal norms. While that may seem indisputable, the fact that an issue of this magnitude could not immediately be relegated to cultural particularism is worth noting in light of discussions between communitarianists and liberals in subsequent decades. More subtle distinctions, such as contextualising the social role of nuclear technology (compare Sweden and Pakistan), can certainly be made. But the hazards of nuclear technology as manifested in weapons, reactor accidents and waste dictate that the ultimate consequences be related to universal norms.

When progress is rejected, the future will be saved by a renunciation of technological progress. German authors such as Jaspers and Jonas are not the only ones who have criticised technological society. The literature of the 1950s and 1960s, which viewed technological progress as a deterministic force unstoppable by all but the most tenacious resistance, addressed this very question. Lewis Mumford in the United States and Jacques Ellul in France (both 1964) see technology as possessing a pervasive logic that has serious consequences for individuals and society. Mumford described the atom bomb as the ultimate proof that technology threatened to annihilate civilisation, and perhaps the human race along with it.[22] Ellul also pointed to nuclear waste as confirmation that technology posed a threat to humanity.[23] Little doubt remained in this literature that technology had created a genuine problem that demanded a solution. Technology is once again regarded as an abstract entity. It takes control of humanity, as Thomas J. Misa (2003) points out in reference to Mumford, Ellul, Heidegger, Horkheimer and Habermas. But Misa argues that such a perspective blurs the distinction between such radically different technologies as hydrogen bombs and contraception. In its modern guise, technological progress is a multifaceted phenomenon.[24] By the same token, nuclear technology is vital to radiology, which has saved many lives, while also generating radioactive waste that cries for a solution.

Progress can, third, be redefined in terms of politics and ethics. Soft Enlightenment ideas take priority over hard ones. Such a posture is wholly reasonable. Given the existence of nuclear waste and the conditions for legitimacy (equity, democracy, and so on), any sustainable solution must be politically and ethically rooted. The next section will analyse the idea of deep geological disposal in light of two definitions of progress.

Deep geological disposal and the idea of progress

Plans for deep geological disposal raise an issue that can be related to the Enlightenment Project and idea of progress, the belief that the future offers infinite potential for new accomplishments. The production of electricity with nuclear power and the future management of its waste products is an application of the Enlightenment Project. Nuclear power and waste management programmes reflect the Enlightenment's unwavering faith in technology and its potential progress. The axiom is that nuclear power is needed to maintain and expand a country's prosperity, and that any problems that arise will be amenable to a solution.

But do these programmes proceed from 'hard' or 'soft' Enlightenment ideas? The hard ideas are those that promote technology, economics and administration as the bulwark of social progress, whereas the soft ideas are those that focus on ethics, religion and democracy. The distinction has been made by Swedish intellectual historian Sven-Eric Liedman in relation to the Enlightenment philosophers and thinking that have shaped Western society for the past 300 years.[25]

Hard Enlightenment ideas were characterised by the type of relentlessness that is visible in nuclear power and waste management programmes. The prospect of technological progress, economic growth and expansion of prosperity is the driving force. The availability of technology constitutes a powerful imperative for progress among engineers, bureaucrats and public officials. From this point of view, legislation and public policy represent a pitfall that must be avoided or speed bumps that must be endured on the way to a predefined destination. A critique of the decisions that led to the construction of nuclear power plants can identify many examples of this mindset. Initial optimism was followed by the problem of satisfying demands for examination of the decisions on a more democratic basis. As Sweden closed down some reactors, it expanded the capacity of others, the result being that it has the capacity to produce more nuclear power than ever before.

Typical of soft Enlightenment ideas were ethical introspection, conscious values and deliberation about how to behave in various situations. Official inquiries and reviews before passing laws, as well as judicial proceedings that consider applications for licences to manage nuclear waste based on that legislation, reflect such ideas. Consultation and dialogue with local residents and political representatives are also an outgrowth of soft Enlightenment ideas. Ethical considerations are brought up sporadically, probably because the principle that each generation and country is responsible for its own nuclear waste has been so well established for the last couple of decades.

Hard and soft Enlightenment ideas – technological or economic progress on the one hand and political or ethical evolution on the other – clearly harbour an inherent tension.[26] The concept of deep geological disposal borrows from both sets of ideas in its attempt to provide both a technological and an ethically sustainable solution that bridges the gap between constant scientific advances and the dictum that future generations must not inherit the nuclear waste of those that preceded them. Neither the triumphs of military technology nor the rapid consumption of fossil fuel that is changing the world's climate proceeded from such an imperative.

Both the economics and the technology of nuclear waste management are now global phenomena. Some international bodies and treaties promote soft Enlightenment ideas. However, the International Atomic Energy Agency (IAEA), the Nuclear Energy Agency (NEA) and Euroatom all advocate for the peaceful use of nuclear technology and energy. They cannot deal with the challenges posed by the international scope of waste management technology and economics. Democratic and ethical considerations are dealt with at the national level instead. The hard ideas of the Enlightenment have been internationalised while the soft ideas are limited to the national stage. The dichotomy may soon become so great that democratic and ethical considerations lose much of their influence on decision-making processes.

Such developments would be in stark contrast to that which is urgently needed: a sense of responsibility that proceeds from the human condition in a more fundamental way than the defence of political power, special interests and unwarranted hope can offer.

Enlightenment thought views the future as a blank slate or a range of possibilities. The secularisation process brought the blank slate down from heaven to earth. Paradise was no longer the domain of the hereafter but the legitimate goal of governments that wanted to win the trust of their citizens. Prospects for the peaceful use of nuclear energy aroused enormous enthusiasm in the late 1950s and early 1960s. The construction of reactors contributed to the spirit of modernisation in France. The label of a wine cooperative in Chinon featured lush vine branches with the new reactors in the background.[27]

The immediate source of the deep geological disposal concept is the eagerness of the late 1950s and early 1960s, tamed by the scepticism and criticism that gradually arose. The nuclear waste issue that first emerged in the 1970s broadened and deepened the ideological discourse about the positive and negative implications of modernity. When French architect Claude Parent was recently asked about having agreed to design the buildings for the nuclear power plants in Cattenom and Chooz, he responded in a utopian, futuristic spirit:

> For me, atomic energy was the ultimate expression of speed and power. Giving it form and bridling it was an enormous challenge. Anything related to the atom and particle acceleration was automatically good in my eyes. It wore the emblem of greatness and modernity. Nuclear waste was as foreign a concept as exhaust. That's all there was to it.[28]

From that point of view, deep geological disposal is a kind of compromise between an affirmation and a denunciation of modernity and progress. A 1976 critique by the Swedish Society for Nature Conservation expressed scepticism about the possibility of devising technological solutions that could fully deal with the problem of nuclear waste:

> Some elements must be stored for thousands or hundreds of thousands of years in order for their radioactivity to decay. Certain expensive measures can probably reduce the quantities of the longest-lived elements. However, guarantees must be found to ensure that radioactive waste does not leak into the environment during epochs that are unfathomable from a human perspective.[29]

Such a formulation combines objections to nuclear energy with the concession that it may be possible to curb its hazards. The concept of deep geological disposal is a response to that challenge. Not surprisingly, its origins roughly coincide with the report of the Brundtland Commission concerning the need for sustainable economic growth.

The concept of deep geological disposal creates a discursive field that links faith in progress to a critique of modernity. Nuclear energy is accepted, but its negative consequences must be banished from the face of the earth. The concept makes it easier to defend nuclear energy production. In terms of general arguments, proponents of nuclear energy can answer critics of modernity by pointing to the technological progress that will make deep geological disposal possible. If nuclear energy is denounced as having been rendered obsolete by solar, wind and hydropower, deep geological disposal can still be cited as a reason to continue producing it. In terms of specific arguments, deep geological disposal can help promote nuclear energy as one of the ways to supply the world with sufficient energy as fossil fuels are phased out. The marriage of nuclear energy to deep geological disposal holds out the promise of avoiding the environmental hazards posed by fossil fuels.

That the concept of deep geological disposal borders tenuously on the Enlightenment Project or modernity in general becomes even more obvious when the term final disposal is used. Final disposal proceeds from the assumption that no conceivable technology can either neutralise or radically mitigate the hazards of nuclear waste. While it may leave the door open for new methods of recovering the waste for productive use, there is an undertone of pessimism that future generations need to be protected from insuperable problems. Thus, deep geological disposal speaks the language not only of the Enlightenment, but of Oswald Spengler: civilizations are born, grow, flourish, decline and eventually die.[30]

The concept of deep geological disposal includes pessimism about progress. 'Final disposal' rejects the idea that technological advances will permit the neutralisation of nuclear waste. A future is portrayed in which our civilisation has fallen, technology has regressed and humanity is defenceless against the hazards of nuclear waste. Such declines have occurred throughout history, though limited to particular regions and spheres of knowledge. But there is little evidence that such a collapse is in the offing. In this sense, the concept of retrievability or reversibility is less pessimistic; technical solutions may be found, alternative options for management may one day be developed that finally resolve the issue. In the long run, progress might offer salvation. Nevertheless, when considered from the time horizon of one or two centuries, retrievability appears to be pessimistic.

The concept of deep geological disposal (final disposal and retrievability) is also pessimistic in the sense that it insists that waste must be buried immediately. The argument is that society of the future cannot be trusted to make the right decisions. This is illustrated by the president of Swedish SKB, Claes Thegerström:

> We know today that the rock will be the same in 100 or even 10,000 years. But what about society? Will we still have today's resources, expertise and commitment in 100 years? The more we delay acting, the greater the societal uncertainties we have to contend with.[31]

The underlying message of deep geological disposal is that economic and technological progress on the one hand and democratic and ethical progress on the other are both highly questionable going forward.

The concept of deep geological disposal implies a flight from Enlightenment ideas of progress and its approach to problems, projects and dilemmas as worldly tasks that can be undertaken in the spirit of human solidarity. The concept draws from mythological and religious connotations. It directs our attention to a time horizon that corresponds to the eternity that Christianity posits after earthly life. The time perspective is so vast that human society and historical eras are no longer meaningful constructs. The context becomes religious instead.

Tolkien's *Lord of the Rings* is often cited in reference to nuclear technology and waste. The ring was forged long ago; it possesses miraculous powers but can be used only for Sauron's nefarious purposes. If he can arrogate it to himself, nothing can stop his conquest of the world. The only salvation is to return it to the volcano were it was made, to be destroyed in the lava flows. That is about as good an allegory of deep geological disposal as you could imagine. Uranium mineral has been mined, uranium has been extracted and turned into fuel pellets – now the idea is to restore it to the depths of the earth.

Tolkien's tale also bears witness to the dream of recovering the original harmony of humanity and nature. His is a retelling of an ancient myth with religious undertones. The human race first lives in a veritable paradise, which is subsequently shattered by a calamitous event. We have now fallen from grace, but there is hope that the primal unity can be re-established.

The concept of deep geological disposal points in the wrong direction. The hope of restoring some kind of original harmony between humanity and nature that has been shattered by evil forces is an illusion. Splintered community, a theme that has accompanied progress and modern social development, is here to stay. Goethe's Faust describes the straits in which humanity finds itself in the scene 'Before the Gate':

> One passion only has thy heart possessed;
> The other, friend, O, learn it never!
> Two souls, alas! are lodged in my wild breast,
> Which evermore opposing ways endeavour,
> The one lives only on the joys of time,
> Still to the world with clamp-like organs clinging;
> The other leaves this earthly dust and slime,
> To fields of sainted sires up-springing.[32]

Notes

1 Immanuel Kant *Die Metaphysik der Sitten*, Frankfurt am Main: Suhrkamp Verlag, 1956 and 1977 (1797), pp. 522f.
2 Giorgio Agamben *Undantagstillståndet*, Site: Lund 2005 (2003), p. 65, 2005.
3 Tore Nordenstam *The Power of Example*, Stockholm: Santéreus Academic Press, 2009, pp. 227–32.
4 Karl-Otto Apel *Diskurs und Verantwortung: Das Problem des Übergangs zur postkonventionellen Moral*, Frankfurt am Main: Suhrkamp 1990 (1988), pp. 163ff.
5 Apel op. cit. pp. 141–49.
6 Ulrich Beck *Risikogesellschaft: auf dem Weg in eine andere Moderne*, Frankfurt am Main 1986. Ulrich Beck and Edgar Grande *Das kosmopolitische Europa: Gesellschaft und Politik in der Zweiten Moderne*. Frankfurt am Main: Suhrkamp Verlag, 2004.

7 Jürgen Habermas *The Postnational Constellation*, Cambridge, MA: MIT Press, 2001 (1998).
8 John Urry *Global Complexity*, Cambridge: Polity Press, 2003.
9 Urban Strandberg and Mats Andrén 'Editorial: Nuclear waste management in a globalised world', *Journal of Risk Research* 7–8/2009.
10 Mats Andrén *Mellan deltagande och uteslutning: det lokala medborgarskapets dilemma*, Hedemora: Gidlunds förlag, 2005. *Local Citizenship, Göteborg* 2007, *Den europeiska blicken och det lokala självstyrets värden*, Hedemora: Gidlunds förlag, 2007.
11 Strandberg and Andrén op. cit
12 Ibid.
13 Jürgen Habermas *Die neue Unübersichtlichkeit*, Frankfurt am Main: Suhrkamp Verlag, 1985, pp. 141ff.
14 Helmut Fleischer *Ethik ohne Imperativ: zur Kritik des moralischen Bewusstseins*, Frankfurt am Main: Fischer Verlag, 1987, pp. 236ff.
15 Habermas op. cit. pp. 80–90.
16 Ulrich Beck *Den kosmopolitiska blicken*, Göteborg: Daidalos, 2005 (2004), citation p. 25.
17 Peter Sloterdijk 'Fern-Nachbarschaft' in *Die Zeit* 26 April 2007, p. 9. 'Mann muss nicht mehr zusammenleben, um zu sein Verbund, mann muss nicht sein verwandt, um etwas füreinander übrigzuhaben; mann muss keine illusion Nahr gemeinsam, um sich miteinander zu Solidarisieren; mann muss sich nicht gesehen haben persöhnlich, um etwas zu tun füreinander. Ich nenne die Summe dieser Verhältnisse: "Fern-nachbarschaften." "Ich Nenne die Summe dieser Verhältnisse: Fern-nachbarschaften."'
18 Jürgen Habermas *The Postnational Constellation*, Cambridge, MA: MIT Press, 2001 (1998).
19 Hans Jonas *Das Prinzip Verantwortung: Versuch einer Ethik für die technologische Zivilisation*, Frankfurt am Main: Suhrkamp 1984 (1979). See Chapter 5.
20 Johan Hendrik Jacob van der Pot meticulously examined various views of technological progress in *Die Bewertung des technischen Fortschritts: Eine systematische Übersicht der Theorien*, Maastricht: von Gorcum, 1985.
21 Andrew, Blowers 'Editorial: Why Fukushima is a moral issue? The need for an ethic for the future in the debate about the future of nuclear energy' *Journal of Integrative Environmental Sciences* 2/2011, p. 79.
22 Lewis Mumford *The Pentagon of Power*, San Diego: Harcourt Brace Jovanovich 1970 (1964), pp. 230–37.
23 Jacques Ellul *The Technological Society*, New York: cop., 1964 (1954).
24 Thomas J. Misa 'The Compelling Tangle of Modernity and Technology' in Thomas J. Misa, Philip Brey and Andrew Feenberg (eds) *Modernity and Technology*, Cambridge MA: MIT Press, 2003.
25 Sven-Eric Liedman *I skuggan av framtiden*, Stockholm: Bonnier, 1997.
26 Ibid.
27 Gabrielle Hecht *The Radiance of France: Nuclear Power and National identity after World War II*, London: MIT Press, 2009 (1998), pp. 261ff.
28 Niklas Maak 'Auf der schiefen Bahn' in *Frankfurter Allgemeine Sonntagszeitung* 1 August 2010. Citat: 'Es gab damals für mich nichts Schnelleres, Kraftvolleres als das Atom. Es war eine Riesenaufgabe, diese Energie in Form zu bringen, zu bändigen. Ich fand alles, was mit Atom und Teilchenbeschleunigung zu tun hatte, gut. Es war gross und modern. Über Atommüll hatte man damals so wenig nachgedacht wie über Abgase. So einfach war das.'
29 Swedish Society for Nature Conservation 'Energifrågan-naturvårdssynpunkter', 1976, p. 12: 'Vissa ämnen måste lagras i tusentals eller hundratusentals år för att radioaktiviteten skall kunna avklinga. Genom vissa fördyrande åtgärder kan man

troligen minska mängderna av de mest långlivade ämnena. Under dessa från mänsklig synpunkt mycket långa tidsperioderna måste dock garantier finnas för att inte radioaktivt avfall skall läcka ut i omgivningen.'

30 Oswald Spengler *Der Untergang des Abendlandes*, München: C. H. Beck'sche, 1918–22.

31 Cited from Malcolm Grimston 'Ethical and environmental principles', Committee on Radioactive Waste Management/CORWM 2004.

32 Goethe, Johann Wolfgang von Faust, Köln 2005 (1813): 'Du bist dir nur des einen Triebs bewusst;/O lernen nie den anderen kennen!/ Zwei Seelen wohnen, ach! In meiner Brust,/ Die eine will sich von der anderen trennen;/ Die eine hält in derber Liebeslust/ Sich an die Welt mit klammernden Organen;/ Die andre hebt gewaltsam sich von Dust/ Zu den Gefilden hoher Ahnen.'

Conclusion

Legitimacy and ethics

In this book it is argued that neither fully adequate nor complete legitimacy is possible. It might be argued that the various meanings of the concept can be integrated such that complete legitimacy can be achieved by satisfying all of them. For a number of reasons, such an amalgam is not feasible. Alexis de Tocqueville and John Stuart Mill both discuss the conflict inherent to modern society between individual liberty and the growing power of the state. Habermas speaks of a crisis of legitimacy created by the welfare state when it assumes various commitments but is unable to keep its promises. Inherent to political power is that it must be established legitimately while withstanding shattered expectations and scepticism about what it has accomplished. Thus, legitimacy in modern society is always in crisis, never more than when it comes to nuclear waste management.

Another general conclusion can be drawn from the observation that the concept of legitimacy is always in flux. New, unpredictable meanings emerge. Changes in the various constellations that make up the social order also affect the concept. The demands to which legitimacy is subject will certainly evolve as time goes on. The process will inevitably continue, but it is impossible to foresee its contours.

The history of nuclear waste management illustrates the way that demands for legitimacy arise, generate new norms, and pose the following dilemma: legitimacy must be established if anything is to be accomplished, but such attempts suffer under the yoke of never-ending crisis. The dismaying conclusion is that nuclear waste management cannot be fully legitimised. However, it can attain more or less legitimacy.

A distinction must be made between what can be solved permanently and what cannot. The radiation generated by nuclear waste requires a solution that permanently isolates it from the biosphere. The legitimacy of nuclear waste management poses a dilemma that operates on another level, where it cannot be resolved once and for all. The emphasis must be on learning to live with nuclear waste as wisely as possible. The dilemma cannot be expected to go away in the foreseeable future.

The dilemma of nuclear waste has implications for a fundamental principle of what constitutes a democratic society. The problem of nuclear waste challenges the maxim that citizens both participate in and comply with public policy decisions. Any solutions that are implemented today restrict the options of future generations. In addition, the legitimacy that is created must be sustainable over the long term. But how can that be accomplished? Legitimacy criteria can change during the decades required to build a repository, allow the waste to cool, encapsulate it and bury it.

The inevitable conclusion is that it is impossible to create the kind of legitimacy that is equal to the long-term challenge of the nuclear waste issue. No final solution can be regarded as legitimate at this point. The only reasonable approach is to wait and further develop various alternative solutions. Keep the waste under surveillance and strictly isolated from the biosphere. Let it generate momentum to identify better solutions than those currently available. Let proposals for deep geological disposal meet the same fate as plans to bury nuclear waste under Antarctic ice caps or send it to the moon.

Nuclear plants are the fulfilment of dreams about large-scale energy production. But the waste they generate has turned out to be a nightmare. The hazards force humanity to dream about common, transnational norms. Nuclear energy and waste create an arena that weaves together different interests, illustrating the cosmopolitisation that Beck speaks of, which brings with it knowledge and sensitivity to conditions in other parts of the world.[1] However, the process has a different impact from country to country. Developing nations see the wealth and success chalked up by nuclear energy producers and want to join the club. One country's dream is another country's nightmare, particularly when nuclear energy is exploited to develop weapons. Nightmare scenarios demand stringent monitoring procedures to prevent the nuclear threshold from being lowered. The peaceful use of nuclear energy requires global collaborative structures and cosmopolitan institutions.

What are the possible contexts for discussion and the formulation of norms? A variety of contexts are needed to ensure that the results are not chained to a specific ideology. Neither neo-liberalism, nationalism nor any other contemporary ideology must be allowed to establish all the norms.

A voluntary international agreement or declaration of cosmopolitan norms is needed to deal with the entire nuclear fuel cycle. The agreement should specify an ethical framework for the formulation of norms. The interests of both current and future generations must be taken into consideration. The norms do not have to be so general that they are immune to social change. Individual safety and security should be given top priority. Private interests cannot be allowed to stand in the way of the common good. Risks to future generations must be eliminated.

One option is to set up an ethics committee. This idea, which is fully in line with the Kantian tradition, was first articulated in the 1975 Declaration of Helsinki with respect to biomedical research. According to the declaration, the interests of science and the community are subordinate to the welfare of test subjects.[2] The UN Panel of Climate Scientists is one of many such bodies that address various issues around the world. In the case of the nuclear fuel cycle, a committee or panel could formulate norms and quantify the variables they specify. The committee would be diverse in terms of nationality, ethnicity and gender. Its members would be prominent representatives of the social sciences, humanities, religious communities and literary world as well as leading scientists and engineers. The committee would work with an international court in an advisory capacity.

The court would be the licensing authority for all stages of the nuclear fuel cycle – extraction, production and waste management.

A second option would be to reform the IAEA. A first step would be to make its guidelines binding on member states.

But can the legitimacy of nuclear waste management be strengthened by institutionalising norms for it? Proceeding from Apel's thinking, the institutional approach to creating legitimacy is insufficient because it must also be clearly linked to ethical reason and discourse. Some standards and norms can be institutionalised separately, but the legitimacy thereby achieved may soon lose its vitality.[3] Thus, it is essential that the issues surrounding nuclear waste management be resolved within the framework of ethical discourse.

This is a far cry from relying on an eternal, complacent ethical principle as the ultimate guideline for managing nuclear waste. Any such principle will quickly collapse as times change, or will erode to an empty phrase. It will divert the discussion from issues of equity and democracy. If such a situation is to be avoided, ethical questions must be placed in a broader context.

The focus should be on the relationship between ethics and legitimacy. While a personal ethic or ethic of the Other links nuclear waste management to intragenerational equity and democracy, an impersonal ethic stresses responsibility to future generations. Only a combination of the two can broadly and sustainably relate ethical issues surrounding the good life to the concept of legitimacy. Successfully doing so can have decisive consequences. Legitimacy issues will radicalise ethics by asking questions about equity and democracy. Meanwhile, ethical issues will radicalise the concept of legitimacy by including future generations. Thus, the discussion of ethics has immediate implications for the contemporary use of the legitimacy concept

The relationship between ethics and legitimacy must be regarded as dialectic. On the one hand, ethics is among many ingredients of legitimacy. Nuclear waste management is a broader, more complex issue than establishing a set of ethical assumptions. Bearing in mind that nuclear waste will be around for a hundred years or considerably longer, a reasonable focus would be on the legitimacy issues that arise. Ethics is still a factor to be considered, but alongside legality, popular will, tradition, emotion, values, best science/technology and communication/dialogue. On the other hand, ethics takes precedence over the other factors in the sense that it points to the normative basis for the community within which legitimacy is created.

Thus, nuclear waste management policy faces two immediate tasks. The first task is to craft an institutional framework that can create legitimacy. Both new international institutions and a global legal system need to be established. The second task is to relate nuclear waste management to common, cosmopolitan values that the world community can accept. The nature of these values will depend on the ability of the community to formulate norms.

Notes

1 Beck op. cit., p. 25.
2 Kemp op. cit., pp. 193–201.
3 Apel på. cit., pp. 55f.

Bibliography

Agamben, Giorgio *Undantagstillstånde*, Site: Lund 2005.

Ahearn, John F. 'Intergenerational Issues Regarding Nuclear Power, Nuclear Waste, and Nuclear Weapons' in *Risk Analysis* 6, 2000.

Åkermark, Torbjörn 'Oseriöst bygga slutförvar för kärnavfallet redan nu', DN. Debatt, 31 July 2010.

Andersson, Digby C. *What Has Ethical Investment to Do with Ethics*, 1998.

Andersson-Skog, Lena 'Från en energi till farligt avfall – kärnkraftsfrågans reglering i det svenska välfärdsbyggandet. En ekonomisk historisk översikt' in Andrén and Strandberg (eds) *Kärnavfallets politiska utmaningar*, Hedemora: Gidlunds, 2005.

Andrén, Mats *Borgerskapets Marx: Eugen von Böhm-Bawerk ämbetsman och nationalekonom i sekelskiftets Wien*, Stockholm: Symposion, 1990.

——*När den nya nationalekonomin kom till Sverige: marginalismen, den österrikiska skolan och Knut Wicksell*, Göteborg: Göteborgs universitet, 1994.

——*Mellan deltagande och uteslutning: det lokala medborgarskapets dilemma*, Hedemora: Gidlunds förlag, 2005.

——*Local Citizenship, Göteborg*: Göteborgs universitet, 2007.

——*Den europeiska blicken och det lokala självstyrets värden*, Hedemora: Gidlunds förlag, 2007.

Andrén, Mats and Strandberg, Urban (eds) *Kärnavfallets politiska utmaningar*, Hedemora: Gidlunds förlag, 2005.

Ansell, C. K. 'Legitimacy: Political', in *International Encyclopedia of the Social and Behavioral Sciences* 2002.

Apel, Karl-Otto *Diskurs und Verantwortung: Das Problem des Übergangs zur postkonventionellen Moral*, Frankfurt am Main: Suhrkamp 1990.

Badie, B. 'Legitimacy, Sociology of', in *International Encyclopedia of the Social and Behavioral Sciences* 2002.

Ballard, Kevin R. and Kuhn, Richard G. 'Developing and Testing a Facility Location Model for Canadian Nuclear Fuel Waste' in *Risk Analysis* 6, 1996.

Barthe, Yannick 'Framing nuclear waste as a political issue in France' in *Journal of Risk Research* 7–8/2009.

Basset, Gilbert W.; Jenkins-Smith, Hank C. and Silva, Carol, 'On site storage of High-Level Nuclear Waste: Attitudes and Perceptions of Local Residents' in *Risk Analysis* 3, 1996.

Beck, Ulrich *Risk Society: Towards a New Modernity*, London: Sage 1992.

——*The Cosmopolitan Vision*, Cambridge: Polity, 2006.

——*Den kosmopolitiska blicken*, Göteborg: Daidalos, 2005.

Beck, Ulrich and Grande, Edgar *Das kosmopolitische Europa: Gesellschaft und Politik in der Zweiten Moderne*, Frankfurt am Main: Suhrkamp Verlag, 2004.

Beetham, David *The Legitimation of Power*, Basingstoke: MacMillan 1991.

Beetham, David and Lord, Christopher *Legitimacy and the European Union*, London: Longman, 1998.

Berkhout, Frans *Radioactive Waste: Politics and Technology*, London: Routledge 1991.

Bloch, Ernst *Prinzip Hoffnung*, Frankfurt am Main: Suhrkamp, 1985.

Blowers, Andrew 'Nuclear waste and landscapes of risk', *Landscape Research* 3, 1999.

——'Why Fukushima is a moral issue? The need for an ethic for the future in the debate about the future of nuclear energy', *Journal of Integrative Environmental Sciences* 8, 2011.

Blowers, Andrew; Lowry, David and Solomon, Barry *The International Politics of Nuclear Waste*, London: Croom Helm 1991.

Burke, Peter *The fabrication of Louis XIV*, London: Yale University Press, 1992.

Coudenhove Calergi, Richard N. *Revolution durch Technik*, Wien: Paneuropa Verlag, 1932.

Cramér, Per; Erhag, Thomas and Stendahl, Sara *Nationellt ansvar för använt kärnbränsle*, Stockholm, Santérus Academic Press, 2009.

Dawson, Jane I. and Darst, Robert G. 'Russia's proposal for a global waste repository: safe, secure and environmentally just?' *Environment* 47(4), 2005.

Deutsche Wörterbuch von Jacob Grimm und Wilhelm Grimm, Leipzig, 1854–1971.

Dostoevsky, Fjodor *Brothers Karamazov*, Constance Garnett's translation. New York: Random House, 1995.

Durant, Darrin 'Radwaste in Canada: a political-economy of uncertainty' in *Journal of Risk Research* 7–8/2009.

Elam, Mark and Sundqvist, Göran 'The Swedish KBS Project: A Last Word in Nuclear Fuel Safety Prepares to Conquer the World?' in *Journal of Risk Research* 7–8/2009.

——'Meddling in Swedish Success in Nuclear Waste Management' in *Environmental Politics* 2, 2011.

Ellul, Jacques *The Technological Society*, New York: cop., 1964 (1954).

Etymologisches Wörterbuch des Deutschen, Berlin: Akademie Vgl, 1993.

Fish, Stanley *The Trouble with Principle*, London: Harvard University Press, 2001 (1999).

Fleischer, Helmut *Ethik ohne Imperativ: zur Kritik des moralischen Bewusstseins*, Frankfurt am Main: Fischer Verlag, 1987.

Garriga, Elisabet and Mele, Domènec 'Corporate Social Responsibility Theories: Mapping the Territory' in *Journal of Business Ethics* 53, 2004.

Goethe, Johann Wolfgang von *Faust*, Köln 2005 (1813).

Goodin, Robert E. 'Uncertainty as an Excuse for Cheating Our Children: The Case of Nuclear Waste' in *Policy Sciences* 10, 1978.

Grimston, Malcolm 'Ethical and environmental principles', Committee on Radioactive Waste Management Committee on Radioactive Waste Management CORWM 2004.

Gustafson, Gunnar 'De tekniska principerna bakom det svenska slutförvaret för använt kärnbränsle – KBS 3' in Mats Andrén and Urban Strandberg (eds) *Kärnavfallets politiska utmaningar*, Hedemora: Gidlunds förlag, 2005.

Habermas, Jürgen *Die neue Unübersichtlichkeit*, Frankfurt am Main: Suhrkamp Verlag, 1985.

——*Faktizität und Geltung: Beiträge zur Diskurstheorie des Rechts und des demokratischen Rechtsstaats*, Frankfurt am Main: Suhrkamp Verlag, 1992.

——*The Postnational Constellation*, Cambridge, MA: MIT Press, 2001.

——*Den moraliska synpunkten*, Göteborg: Daidalos, 2008.

Häntsch, Carola 'The World Citizen from the Perspective of Alien Reason: Notes on Kant's Category of the *Weltbürger* according to Josef Simon' in Rebecka Lettevall and My Klockar Linder (eds) *The Idea of Kosmopolis: History, philosophy and politics of world citizenship*, Huddinge: Södertörn Högskola 2008.

Hecht, Gabrielle *The Radiance of France: Nuclear Power and National Identity after World War II*, London: MIT Press, 2009.

Heidegger, Martin *Nietzsche: Europäischer Nihilismus*, Gesamtausgabe Bd. 48, Frankfurt am Main: Klostermann, 1986.

Historisches Wörterbuch der Philosophie, Basel: Schwabe, 1971–2007.

Hocke, Peter and Renn, Ortwin 'Concerned Public and the Paralysis of Decision Making. Nuclear Waste Management Policy in Germany' in *Journal of Risk Research* 7 2009.

Högselius, Per 'Spent nuclear fuel policies in historical perspective: an international comparison', *Energy Policy* 37, 2009.

Holland, I. 'Waste not want not? Australia and the politics of high-level nuclear waste', in *Australian Journal of Political Science* 37(2), 2002.

Horkheimer, Max 'Förnuftets slut' in John Burill (ed.) *Kritisk teori – en introduktion*, Göteborg: Daidalos, 1987. First published as 'End of Reason' in *Studies in Philosophy and Social Science* 1941.

Huemer, Lars; Krogh, George von and Roos, Johan 'Knowledge and the concept of trust' in Georg von Krogh, Johan Roos and Dirk Kleine (eds) *Knowing in Firms: Understanding, Managing and Measuring Knowledge*, Sage: London 1998.

IAEA 'The Principle of Radioactive Waste Management', Vienna 1995.

Internet Encyclopedia of Philosophy, 1995.

Jaspers, Karl *Psychologie der Weltanschauung*, Berlin 1922.

——*Die geistige Situation der Zeit*, Berlin: Sammlung Göschen, 1931.

——*Der philosophische Glaube*, München: R. Piper 1948.

——*Die Atombombe und die Zukunft des Menschen*, München: R. Piper Verlag. 1958.

Jonas, Hans 'Gnosticism, Existentialism and Nihilism', in *The Phenomenon of Life: Towards a Philosophical Biology*, Evanston: Northwestern University Press, 2001.

——*Das Prinzip Verantwortung: Versuch einer Ethik für die technologische Zivilisation*, Frankfurt am Main: Suhrkamp, 1984.

Kant, Immanuel *Die Metaphysik der Sitten*, Frankfurt am Main: Suhrkamp Verlag, 1956 and 1977.

KASAM/National Council for Nuclear Waste *Kunskapsläget på kärnavfallsområdet 2007: nu levandes ansvar, framtida generationers frihet*, SOU 2007:38.

Kemp, Peter *Das Unersetzlische – eine Technologieethik*, Berlin: Wichern Verlag, 1992.

——*Citizen of the World: Cosmopolitan Ideals for the 21st Century*, Lancaster: Prometheus Books, 2010.

Lidskog, Rolf and Sundqvist, Göran 'On the right track? Technology, geology and society in Swedish nuclear waste management' in *Journal of Risk Research* 7–8/2004.

Liedman, Sven-Eric *I skuggan av framtiden*, Stockholm: Bonnier,1997.

Linn, Björn 'Kärnfrågor i samhällsplaneringen' in Andrén, Mats and Strandberg, Urban (eds) *Kärnavfallets politiska utmaningar* 2005.

Maak, Niklas 'Auf der schiefen Bahn' in *Frankfurter Allgemeine Sonntagszeitung* 1 August 2010.

MacDonald, Digby and Sharifi-Asl, Samin 'Is Copper Immune to Corrosion When in Contact with Water and Aqueous Solutions?' SSM Report 2011–09.

MacKerron, Gordon and Berkhout, Frans 'Learning to listen: institutional change and legitimation in UK radioactive waste policy' in *Journal of Risk Research* 7–8/ 2009.

Mann, Thomas *Betrachtungen eines Unpolitischen*. Frankfurt am Main: S. Fischer Verlag, 2002, 1983 [1918].

Marcuse, Herbert 'Några samhälleliga konsekvenser av den moderna teknologin' in John Burill (ed.) *Kritisk teori – en introduktion*, Göteborg: Daidalos 1987, pp. 363, 142. First published as 'Some Social Implications of Modern Technology' in *Studies in Philosophy and Social Science* 1941.

McKeon, Richard 'The development and the significance of the concept of responsibility', in McKeon. Zahava M. (ed.) *Freedom and history and other essays: an introduction to the thought of Richard McKeon*, Chicago: University of Chicago Press 1990.

Mill, John Stuart *Considerations on Representative Government*, London 1867.

——*On Liberty*, London 1859.

Mohan, Ram and Aggarwal, Veena 'Spent fuel management in India' in *Journal of Risk Research* 7–8/2009.

Möller, Ulrika *The prospects of security cooperation: a matter of relative gains or recognition? India and nuclear weapons control*, Gothenburg: University of Gothenburg, 2007.

Mumford, Lewis *The Pentagon of Power*, San Diego: Harcourt Brace Jovanovich 1970 (1964).

NEA 'The Environmental and Ethical Basis of Geological Disposal of Long-Lived Radioactive Waste', Paris, 1995.

Okrent D. 'On Intergenerational Equity and its Clash with Intragenerational Equity and the Need for Policies to Guide the Regulation of Disposal of Wastes and Other Activities Posing Very Long-Term Risks' in *Risk Analysis* 5, 1999.

Oxford English Dictionary, Oxford: Clarendon 1986.

Partridge, Ernest *Responsibilities to Future Generations: Environmental Ethics*, Buffalo, NY: Prometheus Books, 1981 (1980).

Partial verdict, 22 March 2006, handed down in Vänersborg, Vänersborg District Court (Environmental Court).

Perera, J. 'China and Sudan want Germany's nuclear waste', *New Scientist* 107, 1991.

Pot, Johan Hendrik Jacob van der *Die Bewertung des technischen Fortschritts: Eine systematische Übersicht der Theorien*, Maastricht: von Gorcum, 1985.

Putnam, Robert *Making Democracy Work. Civic Traditions in Modern Italy* (1993) Princeton: Princeton University Press.

Rauschning, Hermann *Die Revolution des Nihilismus: Kulisse und Wirklichkeit im Dritten Reich*, Zürich: Europa Verlag, 1938.

Riotte, H. 'Stakeholder Issue at OECD/NEA' in JHPS/NEA Symposium, Stakeholder Involvement on Radiation Protection, 2005. See http://www.soc.nii.ac.jp/jhps/s/events/kikaku/stake-sympo/1-Riotte.pdf.

Rochlin, Gene I. 'Nuclear Waste Disposal: Two Social Criteria' in *Science* 1977, 195.

Shrader-Frechette, Kristin *Burying Uncertainty: Risk and the Case against Geological Disposal of Nuclear Waste*, Berkeley: University of California Press, 1993.

——'Duties to Future Generations, Proxy Consent, Intra- and Intergenerational Equity: The Case of Nuclear Waste' in *Risk Analysis* 6, 2000.

Schulz, Hans *Deutsches Fremdwörterbuch*, Berlin 1926.

Sismondi, Simonde de *Études sur les constitutiones des peuples libres*, Bruxelles 1836 (1815).

SKB/ Swedish Nuclear Fuel and Waste Management Co 'RD & D Programme 2007: Programme for research, development and demonstration of methods for the management and disposal of nuclear waste', Stockholm, 2007.

Skeat, Walter W. *Etymological Dictionary of the English Language*, Oxford 1910.

Sloterdijk, Peter 'Fern-Nachbarschaft' in *Die Zeit* 26 April 2007.

Solomon, Barry 'High-level Radioactive Waste Management in the U.S.' in *Journal of Risk Research* 7–8/2009.

Solomon, Barry D; Andrén, Mats and Strandberg, Urban 'Three Decades of Social Science Research on High-Level Nuclear Waste: Achievements and Future Challenges', *Risk, Hazards and Crisis in Public Policy*, 1:4, 2010.

Sovacool, Benjamin K. 'Critically weighing the costs and benefits of a nuclear renaissance', *Journal of Integrative Environmental Sciences*, 7:2, 2010.

Spengler, Oswald *Der Untergang des Abendlandes*, München: C. H. Beck'sche, 1918–22.

Stendahl, Sara *Communicating Justice Providing Legitimacy: the legal practices of Swedish administrative courts in cases regarding sickness case benefit*, Uppsala: Iustis förlag, 2003.

Stenmark, Mikael and Bråkenhielm, Carl Reinhold 'Nuclear Waste, Ethics and Responsibility for Future Generations' in *Nuclear Waste: state-of-the-art reports 2004*, Stockholm, SOU 2004:67.

Stigh, Jimmy 'KASAM och den Baltiska skölden' in *Kärnavfall: tillbakablick och framtidsperspektiv i KASAM:s verksamhet*, Stockholm, SOU 2004:120.

Strandberg, Urban and Andrén, Mats (eds) Special issue: Nuclear Waste Management in a Globalised World, *Journal of Risk Research* 7–8/2009.

——'Editorial: Nuclear Waste Management in a Globalised World' in *Journal of Risk Research* 7–8/2009.

——(eds) *Nuclear Waste Management in a Globalised World*, London: Routledge, 2011.

Svenska akademiens ordbok, Stockholm 1903.

Swedish National Council for Nuclear Waste 'Kunskapsläget på kärnavfallsområdet 2007: nu levandes ansvar, framtida generationers frihe', Stockholm, SOU 2007:38.

Swedish Society for Nature Conservation 'Energifrågan-naturvårdssynpunkter', 1976.

Urry, John *Global Complexity*, Cambridge: Polity Press, 2003.

Vedung Evert 'Det högaktiva kärnavfallets väg till den rikspolitiska dagordningen', in Andrén, Mats and Strandberg, Urban (eds) *Kärnavfallets politiska utmaningar* 2005.

Wallenius, Janne 'Nyttiggörande eller kvittblivning - transmutation eller bara förvaring?', in Andrén, Mats and Strandberg, Urban (eds) *Kärnavfallets politiska utmaningar* 2005.

WNA 'Environmental and Ethical Aspects', http://www.world-nuclear.org/info/Environ mental_Ethical_Aspects_inf04ap5.html. 1995.

Weber, Max *Wirtschaft und Gesellschaft: Grundriss der Verstehende Soziologie*, Tübingen: Mohr Siebeck, 1972 (1921).

Wolin, Richard *Heidegger's Children: Hannah Arendt, Karl Löwith, Hans Jonas, and Herbert Marcuse*, Princeton: Princeton University Press, 2001.

Würtenberger, Thomas 'Legitimität/Legalität' in *Geschichtliche Grundbegriffe*, Stuttgart: Clett Kotta, 1982.

Index